# SHORT-LIVED MOLECULES

# ELLIS HORWOOD SERIES IN INORGANIC CHEMISTRY

*Series Editor:* J. BURGESS, Department of Chemistry, University of Leicester

Inorganic chemistry is a flourishing discipline in its own right and also plays a key role in many areas of organometallic, physical, biological, and industrial chemistry. This series is developed to reflect these various aspects of the subject from all levels of undergraduate teaching into the upper bracket of research.

| | |
|---|---|
| Almond, M.J. | **Short-lived Molecules** |
| Beck, M.T. & Nagypal, I. | **Chemistry of Complex Equilibria** |
| Burgess, J. | **Ions in Solution: Basic Principles of Chemical Interactions** |
| Burgess, J. | **Metal Ions in Solution** |
| Burgess, J. | **Inorganic Solution Chemistry** |
| Cardin, D.J., Lappert, M.F. & Raston, C.L. | **Chemistry of Organo-Zirconium and -Hafnium Compounds** |
| Constable, E.C. | **Metal Ligand Interactions** |
| Cross, R.J. | **Square Planar Complexes: Reaction Mechanisms and Homogeneous Catalysis** |
| Harrison, P.G. | **Tin Oxide Handbook** |
| Hartley, F.R., Burgess, C. & Alcock, R.M. | **Solution Equilibria** |
| Hay, R.W. | **Bioinorganic Chemistry** |
| Hay, R.W. | **Reaction Mechanisms of Metal Complexes** |
| Housecroft, C.E. | **Boranes and Metalloboranes: Structure, Bonding and Reactivity** |
| Lappert, M.F., Sanger, A.R., Srivastava, R.C. & Power, P.P. | **Metal and Metalloid Amides** |
| Lappin, G. | **Redox Mechanisms in Inorganic Chemistry** |
| Maddock, A. | **Mössbauer Spectroscopy** |
| Massey, A.G. | **Main Group Chemistry** |
| McGowan, J. & Mellors, A. | **Molecular Volumes in Chemistry and Biology: Applications Including Partitioning and Toxicity** |
| Romanowski, W. | **Highly Dispersed Metals** |
| Snaith, R. & Edwards, P. | **Lithium and its Compounds: Structures and Applications** |
| Williams, P.A. | **Oxide Zone Geochemistry** |

# SHORT-LIVED MOLECULES

MATTHEW J. ALMOND, B.Sc., D.Phil.
Department of Chemistry
University of Reading

**ELLIS HORWOOD**
NEW YORK   LONDON   TORONTO   SYDNEY   TOKYO   SINGAPORE

First published in 1990 by
**ELLIS HORWOOD LIMITED**
Market Cross House, Cooper Street,
Chichester, West Sussex, PO19 1EB, England

A division of
Simon & Schuster International Group

Typeset in Times by Ellis Horwood Limited
Printed and bound in Great Britain
by The Camelot Press, Southampton

British Library Cataloguing in Publication Data

Almond, Matthew J.
Short-lived molecules.
1. Free radicals
I. Title
54′.2′24
ISBN 0–13–798554–1

Library of Congress Cataloging-in-Publication Data

Almond, Matthew, J. (Matthew John), 1960–
Short-lived molecules / Matthew J. Almond.
p. cm. — (Ellis Horwood series in inorganic chemistry)
ISBN 0–13–798554–1
1. Excited state chemistry.    2. Molecular theory.
I. Title.   II. Series
QD461.5.A46   1989
541.2′2–dc20                                        89–26802
                                                      CIP

# Table of contents

1   INTRODUCTION . . . . . . . . . . . . . . . . . . . . . . . 9

2   EXPERIMENTAL TECHNIQUES
2.1   Introduction . . . . . . . . . . . . . . . . . . . . . . . .14
2.2   Generation of short-lived molecules . . . . . . . . . . . . . .15
2.3   Flow systems . . . . . . . . . . . . . . . . . . . . . . . .18
2.4   Flash photolysis . . . . . . . . . . . . . . . . . . . . . . .20
2.5   Matrix isolation . . . . . . . . . . . . . . . . . . . . . . .22
2.6   Studies in low-temperature solution . . . . . . . . . . . . . .25
2.7   Chemical trapping . . . . . . . . . . . . . . . . . . . . . .25
2.8   Summary . . . . . . . . . . . . . . . . . . . . . . . . . .26
      References . . . . . . . . . . . . . . . . . . . . . . . . .26

3   PHOTOCHEMISTRY OF METAL CARBONYLS
3.1   Introduction . . . . . . . . . . . . . . . . . . . . . . . .28
3.2   Solution studies . . . . . . . . . . . . . . . . . . . . . . .28
3.3   Flash photolysis studies . . . . . . . . . . . . . . . . . . .32
3.4   Experiments in rigid glasses . . . . . . . . . . . . . . . . .33
3.5   Matrix-isolation studies on $M(CO)_5$ ($M=Cr,Mo$ or $W$) molecules . . . . .34
3.6   Matrix-isolation studies on $Fe(CO)_4$ . . . . . . . . . . . . .39
3.7   Matrix-isolation studies on $Mo(CO)_3$ and $Mo(CO)_4$ . . . . . . .41
3.8   Formation of the molecule $Mo(CO)_5N_2$ . . . . . . . . . . . .42
3.9   The photochemistry of $M(CO)_6$ ($M=Cr$, $Mo$ or $W$) molecules in
      matrices containing oxygen . . . . . . . . . . . . . . . . .44
3.10  Reactions of binuclear and trinuclear metal carbonyls . . . . . . .48
3.11  Studies of related systems . . . . . . . . . . . . . . . . . .51
3.12  Studies in liquid noble gas solutions . . . . . . . . . . . . . .53
3.13  Time-resolved infrared spectroscopic studies . . . . . . . . . .56
3.14  Gas phase studies . . . . . . . . . . . . . . . . . . . . . .58
3.15  Summary . . . . . . . . . . . . . . . . . . . . . . . . . .61
      References . . . . . . . . . . . . . . . . . . . . . . . . .61

**4   REACTIVITY OF METAL ATOMS**

4.1     Introduction . . . . . . . . . . . . . . . . . . . . . . . . . . . . . . . . . . . .64
4.2     Formation of metal carbonyls . . . . . . . . . . . . . . . . . . . . . . .64
4.3     Formation of binary metal oxides. . . . . . . . . . . . . . . . . . . . .68
4.4     Formation of alkene and alkyne complexes . . . . . . . . . . . . .72
4.5     C–H and C–C bond activation . . . . . . . . . . . . . . . . . . . . . . .76
4.6     Other insertion reactions . . . . . . . . . . . . . . . . . . . . . . . . . .78
4.7     Formation of free radicals. . . . . . . . . . . . . . . . . . . . . . . . . .81
4.8     Studies of metal dimers . . . . . . . . . . . . . . . . . . . . . . . . . . .82
4.9     Small metal clusters . . . . . . . . . . . . . . . . . . . . . . . . . . . . . .85
4.10   Reactions of metal clusters . . . . . . . . . . . . . . . . . . . . . . . . .87
4.11   Metal vapour synthesis . . . . . . . . . . . . . . . . . . . . . . . . . . . .88
4.12   Summary . . . . . . . . . . . . . . . . . . . . . . . . . . . . . . . . . . . . . .90
        References . . . . . . . . . . . . . . . . . . . . . . . . . . . . . . . . . . . . .90

**5   DIVALENT SILICON CHEMISTRY**

5.1     Introduction . . . . . . . . . . . . . . . . . . . . . . . . . . . . . . . . . . . .93
5.2     Production of silylenes. . . . . . . . . . . . . . . . . . . . . . . . . . . . .94
5.3     Silylene ($SiH_2$). . . . . . . . . . . . . . . . . . . . . . . . . . . . . . . . .97
5.4     Dimethylsilylene ($Me_2Si$) . . . . . . . . . . . . . . . . . . . . . . . .102
5.5     Silicon dihalides (dihalosilylenes, $SiX_2$) . . . . . . . . . . . . .105
5.6     Reactions of silylenes . . . . . . . . . . . . . . . . . . . . . . . . . . . .107
      5.6.1   Insertion reactions of silylenes . . . . . . . . . . . . . .107
      5.6.2   Addition reactions of silylenes . . . . . . . . . . . . . .110
      5.6.3   Abstraction reactions of silylenes. . . . . . . . . . . .114
      5.6.4   Dimerization reactions of silylenes. . . . . . . . . . .114
      5.6.5   Isomerization reactions of silylenes . . . . . . . . . .116
5.7     Summary . . . . . . . . . . . . . . . . . . . . . . . . . . . . . . . . . . . . .118
        References . . . . . . . . . . . . . . . . . . . . . . . . . . . . . . . . . . . .119

**6   ASPECTS OF ORGANIC PHOTOCHEMISTRY**

6.1     Introduction . . . . . . . . . . . . . . . . . . . . . . . . . . . . . . . . . . .121
6.2     Carbenes (methylenes) . . . . . . . . . . . . . . . . . . . . . . . . . . .121
6.3     Decarbonylation reactions . . . . . . . . . . . . . . . . . . . . . . . . .126
6.4     Cyclobutadiene and the search for tetrahedrane . . . . . . . .127
6.5     Summary . . . . . . . . . . . . . . . . . . . . . . . . . . . . . . . . . . . . .130
        References . . . . . . . . . . . . . . . . . . . . . . . . . . . . . . . . . . . .131

**7   HIGH-TEMPERATURE MOLECULES**

7.1     Introduction . . . . . . . . . . . . . . . . . . . . . . . . . . . . . . . . . . .132
7.2     Silicon monoxide . . . . . . . . . . . . . . . . . . . . . . . . . . . . . . .132
7.3     Small aluminium molecules. . . . . . . . . . . . . . . . . . . . . . . .136
7.4     Small phosphorus molecules . . . . . . . . . . . . . . . . . . . . . . .139
7.5     Metal oxysalts . . . . . . . . . . . . . . . . . . . . . . . . . . . . . . . . .141
7.6     Metal oxides and halides . . . . . . . . . . . . . . . . . . . . . . . . . .145
7.7     Reactions of metal porphyrins. . . . . . . . . . . . . . . . . . . . . .148
7.8     Summary . . . . . . . . . . . . . . . . . . . . . . . . . . . . . . . . . . . . .149
        References . . . . . . . . . . . . . . . . . . . . . . . . . . . . . . . . . . . .150

**8  IONS AND RADICALS**

8.1  Introduction . . . . . . . . . . . . . . . . . . . . . . . . . . . . . . . . . . . 153
8.2  Formation of radicals . . . . . . . . . . . . . . . . . . . . . . . . . . . . . 153
8.3  Formation of radical cations . . . . . . . . . . . . . . . . . . . . . . . . . 156
8.4  Halobenzene radical cations . . . . . . . . . . . . . . . . . . . . . . . . . 158
8.5  Ion-neutral reactions . . . . . . . . . . . . . . . . . . . . . . . . . . . . . 160
8.6  Photochemistry of molecular ions . . . . . . . . . . . . . . . . . . . . . . 161
8.7  Plasma discharge systems . . . . . . . . . . . . . . . . . . . . . . . . . . 164
8.8  Summary . . . . . . . . . . . . . . . . . . . . . . . . . . . . . . . . . . . . 167
     References . . . . . . . . . . . . . . . . . . . . . . . . . . . . . . . . . . . 168

**9  ROUTES TO INORGANIC MATERIALS**

9.1  Introduction . . . . . . . . . . . . . . . . . . . . . . . . . . . . . . . . . . 170
9.2  Semiconductors . . . . . . . . . . . . . . . . . . . . . . . . . . . . . . . . 170
9.3  Metal nitrides and oxides . . . . . . . . . . . . . . . . . . . . . . . . . . . 173
9.4  Metal silicides . . . . . . . . . . . . . . . . . . . . . . . . . . . . . . . . . 174
9.5  Laser-writing experiments . . . . . . . . . . . . . . . . . . . . . . . . . . 175
9.6  Polythiazyl . . . . . . . . . . . . . . . . . . . . . . . . . . . . . . . . . . . 176
9.7  Summary . . . . . . . . . . . . . . . . . . . . . . . . . . . . . . . . . . . . 179
     References . . . . . . . . . . . . . . . . . . . . . . . . . . . . . . . . . . . 179

**10  ATMOSPHERIC AND INTERSTELLAR CHEMISTRY**

10.1  Introduction . . . . . . . . . . . . . . . . . . . . . . . . . . . . . . . . . . 181
10.2  Chlorine nitrate, and atmospheric ozone depletion . . . . . . . . . . . 181
10.3  Atmospheric trace gas analysis . . . . . . . . . . . . . . . . . . . . . . . 182
10.4  Interstellar grains . . . . . . . . . . . . . . . . . . . . . . . . . . . . . . . 184
10.5  Disulphur, and the formation of comets . . . . . . . . . . . . . . . . . . 187
10.6  Unidentified infrared emission . . . . . . . . . . . . . . . . . . . . . . . 187
10.7  Summary . . . . . . . . . . . . . . . . . . . . . . . . . . . . . . . . . . . . 190
      References . . . . . . . . . . . . . . . . . . . . . . . . . . . . . . . . . . . 190

**Index** . . . . . . . . . . . . . . . . . . . . . . . . . . . . . . . . . . . . . . . . . 191

# 1

# Introduction

What is meant by a short-lived molecule? The broad definition I have worked to, in this book, is a chemical species, which under ambient conditions (i.e. a temperature around 25°C, and a pressure close to 1 atm) will, for a combination of kinetic and thermodynamic reasons, decay on a timescale ranging from microseconds, or even nanoseconds, to a few minutes. Thus species such as those which are stable in high-temperature vapours, and those which are transient reaction intermediates are included. I have limited myself almost entirely to a discussion of molecules with short-lived ground states, and I have not included a description of the excited states of molecules whose ground states are stable.

This book gives a general survey of the chemistry of short-lived molecules, including high-temperature species, radicals and molecular ions and photochemical intermediates. It provides a general introduction to the methods by which such species are generated and detected, and gives a discussion of their structure and chemical reactivity. While the book should appeal to all those, including postgraduate students, working in areas of chemistry connected to spectroscopy and structure determination of small molecules, and on reaction mechanisms, it is hoped that final-year undergraduates and others with a general chemical background will also find much to interest them.

In writing a book such as this, the major problem is one of space. The title 'short-lived molecules' encompasses an enormous range of material. To keep within a reasonable number of pages, I have therefore concentrated on a few fairly well-defined areas of chemistry. However, throughout the book, the reader will recognize certain themes. Each chapter is self-contained, but there is cross-referencing from one chapter to another allowing the reader to see how different approaches may be made to the same problem.

The book is intended to appeal to a reasonably general chemistry audience, and the chapters follow approximately the style of articles in the journal *Chemistry in Britain*. In other words they are aimed at a chemically qualified reader, but without much of the detail found in more specialized reviews. I have also tried to keep the number of references within reasonable bounds, so that while key points are referenced, no more than about 50 articles are referred to in each chapter. To complement these references, a number of specialized review articles relevant to the material in this book, and appearing within about the last ten years, are listed chapter by chapter at the end of this introductory chapter [1]. Some of the material is presented in a 'historical' manner, showing how new experiments have been performed on some chemical systems, and how ideas about these systems have then been modified.

The book covers four broad areas: (i) experimental techniques, (ii) photochemical intermediates, (iii) high-energy species, and (iv) applications of studies of short-lived molecules.

Chapter 2 gives a brief outline of the experimental methods used to generate and to study short-lived molecules. It is not appropriate to give here a detailed account of the spectroscopic methods used. For such information the reader is referred to recent books by Hollas, and by Ebsworth, Rankin and Cradock [2]. The techniques of flow systems, flash photolysis, and matrix isolation are all discussed, as are the various detection methods which can be employed with each technique.

Chapter 3 consists of an outline description of metal carboxyl photochemistry. It includes both matrix-isolation and solution studies on the intermediates produced in the photochemical reactions of metal carbonyls. This chapter is complemented by the following one, covering the reactions of metal atoms. Many of the molecules — such as unsaturated metal carbonyls — whose formation by photochemical 'bond-breaking' reactions is discussed in Chapter 3, reappear in Chapter 4, where their production by 'bond-forming' reactions between metal atoms and appropriate ligands — such as carbon monoxide — is described.

The theme changes somewhat in Chapter 5, where a description of the spectroscopy and chemical reactions of divalent silicon compounds (silylenes) is given. However, the following chapter, which gives a very brief account of some aspects of organic photochemistry, ties together much of the information in the preceding three chapters. In particular, divalent carbon compounds, carbenes, bear an obvious resemblance to silylenes. But they are also connected by the isolobal relationship to some of the unsaturated metal carbonyl fragments, which are discussed in Chapters 3 and 4.

While Chapters 3 to 6 are broadly concerned with photochemical reactions, attention in Chapters 7 and 8 turns to high-energy gaseous species. Both high-temperature molecules, and ions and radicals are discussed. Methods of generation, detection and structure determination are given, alongside a description of some of the chemical reactions of such species.

The last two Chapters (9 and 10) look at applications of studies of short-lived molecules. Two areas are covered: the formation of inorganic materials, the discussion being limited almost exclusively to the technique of chemical vapour deposition, and atmospheric and interstellar chemistry. Though apparently widely separated, these are both areas of chemistry where a knowledge of short-lived intermediates is proving to be of great value.

Last I would like to record some apologies and thanks. My apologies go to all those whose work has not been included in this volume. My selection of material for inclusion has often, by necessity, been somewhat arbitrary. However, I have tried to produce a balanced work, without allowing it to become too long. My thanks are due to all those who have helped in any way with the production of this book. In particular, I would like to thank my colleagues at Reading, Robin Walsh and David Rice, for their helpful comments and suggestions, and Miss Rachel Orrin for her help in proof-reading.

## REFERENCES
[1] The following review articles, written during the last ten years, provide useful background reading. They are listed chapter by chapter.

### Chapter2
'Matrix Isolation', R. N. Perutz, *Ann. Rep. Prog. Chem., C, Royal Society of Chemistry* (1985), **82**, 157.
'Raman Studies of Molecules in Matrices', A. J. Downs and M. Hawkins, *Advances in Infrared and Raman Spectroscopy*. (R. J. H. Clark and R. E. Hester, eds); John Wiley: Chichester (1983), vol. 10, p. 1.
'Kinetics of Reactions in Solution: Part II, Fast Reactions', J. E. Crooks, *Ann. Rep. Prog. Chem., C, Royal Society of Chemistry* (1982), **79**, 41.
'Infrared Laser Spectroscopy of Transient Atoms and Molecules', P. B. Davies, *Ann. Rep. Prog. Chem., C, Royal Society of Chemistry* (1987), **84**, 43.
'Techniques for the Kinetic Study of Fast Reactions in Solution', H. Krüger, *Chem. Soc. Rev.* (1982), **11**, 227.
'Contributions of Pulse Radiolysis to Chemistry', J. H. Baxendale and M. A. J. Rodgers, *Chem. Soc. Rev.* (1978), **7**, 235.
'Flash Photolysis Electron Spin Resonance', K. A. McLauchlan and D. G. Stevens, *Acc. Chem. Res.* (1988), **21**, 54.
'Chemical Applications of Extended X-ray Absorption Fine Structure (EXAFS) Spectroscopy', B.-K. Teo *Acc. Chem. Res.* (1980), **13**, 412.
'Matrix Isolation: from Stained Glass Windows to Interstellar Grains', M. J. Almond, *Chem. Br.* (1987), **23**, 533.

### Chapter 3
'Fe(CO)₄', M. Poliakoff and E. Weitz, *Acc. Chem. Res.* (1987), **20**, 408.
'Fe(CO)₄', M. Poliakoff, *Chem. Soc. Rev.* (1978), **7**, 527.
'Metallocenes as Reaction Intermediates', P. Grebenik, R. Grinter and R. N. Perutz, *Chem Soc. Rev.* (1988), **17**, 453.
'Photochemistry of Small Molecules in Low-temperature matrices', R., N. Perutz, *Chem. Rev.* (1985), **85**, 97.
'Aspects of Inorganic Photochemistry', A. Harriman, *Ann. Rep. Prog. Chem., C, Royal Society of Chemistry* (1986), **83**, 3.
*Organometallic Photochemistry*, G. L. Geoffrey and M. S. Wrighton; Academic Press: New York (1979).

### Chapter 4
'Clusters of Transition Metal Atoms', M. D. Morse, *Chem. Rev.* (1986), **86**, 1049.
'Theoretical Aspects of Metal Atom Clusters', J. Koutecky and P. Fantucci, *Chem. Rev.* (1986), **86**, 539.

'Metal Cluster Complexes and Heterogeneous Catalysis', M. Moskovits, *Acc. Chem. Res.* (1980), **13**, 412.

'Alkynes and Metal Atoms', R. W. Zoellner and K. J. Klabunde, *Chem. Rev.* (1984), **84**, 545.

'Photochemical Reactions Involving Matrix-isolated Atoms', R. N. Perutz, *Chem. Rev.* (1985), **85**, 77.


**Chapter 5**

'Laser Photolysis of Silylene Precursors', P. P. Gaspar, D. Holten and S. Konieczny, *Acc. Chem. Res.* (1987), **20**, 329.

'Reactions of Silicon Intermediates', I. M. T. Davidson, *Ann. Rep. Prog., Chem., C, Royal Society of Chemistry* (1985), **82**, 47.

'Multiple Bonding to Silicon', G. Raabe and J. Michl, *Chem. Rev.* (1985), **85**, 419.

'The Silicon–Carbon Double Bond', H. F. Schaeffer III, *Acc. Chem. Res.* (1982), **15**, 283.

'The Silicon–Carbon Double Bond', L. E. Gusel'nikov and N. S. Nametkin, *Chem. Rev.* (1979), **79**, 529.


**Chapter 6**

'The Matrix Isolation Technique and its Application to Organic Chemistry', I. R. Dunkin, *Chem. Soc. Rev.* (1980), **9**, 1.

'Flash Photolysis Studies of Carbenes and their Reaction Kinetics', D. Griller, A. S. Nazran and J. S. Scaiano, *Acc. Chem. Res.* (1984), **17**, 283.

'Laser Flash Photolysis of Some 1,4-Biradicals', J. C. Scaiano, *Acc. Chem. Res.* (1982), **15**, 252.

'Organic Photochemistry', W. M. Horspool, *Ann. Rep. Prog. Chem., C, Royal Society of Chemistry* (1983), **80**, 87.

'Two Photon Laser Excitation in Organic Photochemistry', J. C. Scaiano, L. J. Johnston, W. G. McGimpsey and D. Weir, *Acc. Chem. Res.* (1988), **21**, 22.


**Chapter 7**

'The Characterisation of High-temperature Molecules using Matrix Isolation and Vibrational Spectroscopy', J. S. Ogden, *NATO Adv. Study Inst. Ser., Ser. C* (1981), **C76**, 207.

'Matrix Isolation Studies on Some Molecular Oxoanion Salt Vapours', J. S. Ogden, *Ber. Bunsenges. Phys. Chem.* (1982), **86**, 832.


**Chapter 8**

'Spectroscopy of Molecular Ions in Noble Gas Matrices', L. Andrews, *Advances in Infrared and Raman Spectroscopy.* (R. J. H. Clark and R. E. Hester, eds); John Wiley: Chichester, vol. 7, p. 59.

'ESR Investigation of Molecular Cation Radicals in Neon Matrices', L. B. Knight, Jr., *Acc. Chem. Res.* (1986), **19**, 313.

'Radical Cations in Condensed Phases', M. C. R. Symons, *Chem. Soc. Rev.* (1982), **11**, 227.

'Organic Radical Ions in Rigid Systems', T. Shida, E. Haselbach and T. Bally, *Acc. Chem. Res.* (1984), **17**, 180.

'Novel Radical Anions in the Solid State', F. Williams and E. D. Sprague, *Acc. Chem. Res.* (1982), **15**, 408.

'Organic Plasma Chemistry', L. L. Miller, *Acc. Chem. Res.* (1983), **16**, 194.

'Hydrocarbon Radical Cations', J. L. Courtneidge and A. G. Davies, *Acc. Chem. Res.* (1987), **20**, 90.

'Boron Subhalides', A. G. Massey, *Chem. Br.* (1980), **16**, 588.

**Chapter 9**

'Metallo-organic Compounds', R. H. Moss, *Chem Br.* (1983), **19**, 733.

'Photo-epitaxy, a New Light on Inorganic Materials', M. J. Almond, D. A. Rice and C. A. Yates, *Chem. Br.* (1988), **24**, 1130.

'Metal–Organic Chemical Vapour Deposition', S. J. C. Irvine and J. B. Mullin, *Chemtronics* (1987), **2** 54.

'Polysulphur Nitride — a Metallic Superconducting Polymer', M. M. Labes, P. Love and L. F. Nichols, *Chem. Rev.* (1979), **79**, 1.

**Chapter 10**

'Interstellar Problems and Matrix Solutions', L. J. Allamandola, *J. Mol. Struct.* (1987), **157**, 255.

'Semistable Molecules in the Laboratory and in Space', H. W. Kroto, *Chem. Soc. Rev.* (1982), **11**, 435.

'Ion Molecule Reactions in the Formation of Simple Organic Molecules in Interstellar Clouds and Planetary Atmospheres', W. T. Huntress, Jr., *Chem. Soc. Rev.* (1977), **6**, 295.

'Absorption Bands in the Spectra of Stars, a Crystal Field Approach', P. G. Manning, *Chem. Soc. Rev.* (1976), **5**, 233.

'Theoretical Studies of Interstellar Radicals and Ions, S. Wilson, *Chem. Rev.* (1980), **80**, 263.

'Atmospheric Chemistry Involving Electronically Excited Oxygen Atoms' J. R. Wiesenfeld, *Acc. Chem. Res.* (1982), **15**, 110.

'Laser Magnetic Resonance Spectroscopy and its Application to Atmospheric Chemistry', B. A. Thrush, *Acc. Chem. Res.* (1981), **14**, 416.

'Chemical Reactions and the Nature of Comets', M. Oppenheimer, *Acc. Chem. Res.* (1980), **13**, 378.

[2] The following two books provide background detail to the spectroscopic methods discussed in this work.

*Modern Spectroscopy*, J. M. Hollas; John Wiley: Chichester (1988).

*Structural Methods in Inorganic Chemistry*, E. A. V. Ebsworth, D. W. H. Rankin and S. Cradock; Blackwell Scientific Publications: Oxford (1987).

# 2

# Experimental techniques

## 2.1  INTRODUCTION

It is only within the past sixty or so years that direct observation of short-lived molecules has been possible. During this time great advances have been made both in methods of generating short-lived species, and, more importantly, in methods for their detection. Nowadays studies of such transient species are relatively commonplace. However, this is an area of chemistry where theory, for example in the form of a proposed reaction mechanism, often proceeds in advance of experiment. Postulated reaction intermediates have sometimes remained unknown for many decades before suitable means for their generation and detection have become available. An example is provided by the well-known ozonolysis reaction of alkenes. Many years ago, Criegee [1] proposed that such reactions proceed via carbonyl oxide derivatives (1). However it is only within the last six years that molecules of this type have been spectroscopically characterized by means of the matrix-isolation technique [2].

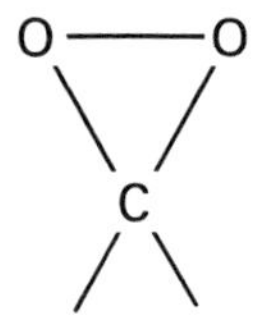

$$O\!\!-\!\!O$$

(1)

Methods to study short-lived molecules can be loosely divided into four categories: flow techniques; flash photolysis and its variations; matrix isolation; and low-temperature solution chemistry. The oldest of these approaches is the flow method, developed independently in the 1920s by Paneth, for demonstrating the existence and reactivity of gas phase free radicals, and by Hartridge and Roughton, for studying biochemical liquid phase processes. Flash photolysis was first reported by

Norrish and Porter in 1949, while matrix isolation, an invention of George Pimentel, was first used in 1953. Studies of chemical reactions in low-temperature solution are assuming an increasing importance as new solvents such as liquid xenon have become available.

More will be said about these techniques, and about the spectroscopic methods of detection of short-lived molecules in the following sections. Here it will suffice to say that all four methods are used today, and that advances in spectroscopic detection have allowed them to be applied to an increasingly wide range of chemical systems. These advances can be partly attributed to the development of reliable and relatively cheap spectrometers, including Fourier transform machines in many areas such as infrared, NMR and, more recently, Raman spectroscopy; to the use of tunable lasers as spectroscopic sources for many experiments; and also to the development of entirely new forms of spectroscopy, such as extended X-ray absorption fine structure (EXAFS) spectroscopy. This technique relies on a source of monochromatic tunable X-rays from a synchroton. Such sources have themselves only recently been designed and commissioned.

## 2.2  GENERATION OF SHORT-LIVED MOLECULES

In this section we will consider the formation of transient intermediates in two distinct types of system. First in high-energy systems, such as high-temperature vapours or plasmas, and second as intermediates of photochemical reactions.

Species in high-temperature vapours are often produced by simple evaporation. Thus molecular alkali metal chlorates ($MClO_3$) [3] or perchlorates ($MClO_4$) [4] are formed by evaporating the corresponding salt at high temperature. There are very many examples of this type of process, and some will be discussed in later sections of this book (see Chapter 7). Sometimes, however, the chemistry of high-temperature vapours is somewhat more complicated. Evaporation of silica ($SiO_2$), for example, yields SiO rather than $SiO_2$ molecules in the vapour phase [5]. This observation reflects the greater thermodynamic stability of the monoxide over the dioxide, at the temperature (ca. 2000 K) at which vaporization takes place. Many other high-temperature reactions likewise involve redox chemistry. Evaporation of sodium phosphate ($Na_3PO_4$ or $Na_5P_3O_{10}$) at ca. 15000 K from a molybdenum boat yields sodium phosphite, $NaPO_2$, in the vapour, since the molybdenum metal acts as a reducing agent [6]. Reduction of phosphorus in the +5 oxidation state is also observed if $OPCl_3$ vapour is passed over silver heated to 1100 K [7]. Here the vapour phase product is OPCl, a molecule which shows an unusual two-coordinate trivalent phosphorus atom, and which has been characterized by a combination of matrix-isolation infrared spectroscopy and mass spectrometry of the vapour.

Among the other methods which may be used for generating 'high-temperature' gaseous species there are three which are of quite widespread use and which we will consider here. These are sputtering, the use of shock waves, and flames. Sputtering is a process in which energetic particles are used to bombard a target and physically to eject atoms from the target surface. For many applications argon ($Ar^+$) ions are used and are accelerated towards the surface with a typical potential of 0.5–4 kV. Fig. 2.1 illustrates the mode of interaction of an $Ar^+$ ion with a metal surface. In this way vapour phase metal atoms are produced, and their reactions may be investigated.

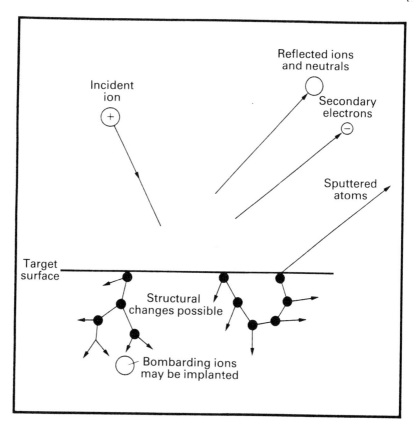

Fig. 2.1 — Sputtering — the interaction of ions with surfaces. Reproduced with permission from Simpson. *Chem. Br.*, **22**, 733 ©1986 Royal Society of Chemistry.

For example, the molecule $WO_2$ has been made by the reaction of sputtered tungsten atoms with $O_2$ [8].

Shock waves can be generated in a number of ways, but the method of greatest importance for studying reactions of short-lived fragments is the shock tube. The basic experimental layout for such a piece of apparatus is shown in Fig. 2.2. A high pressure of gas is held behind a diaphragm, so that the diaphragm is almost at bursting point. By piercing the diaphragm a shock wave is released into the experimental section of the apparatus, which has been held at low pressure. A driver pressure of 3 atm combined with a gas pressure in the experimental section of about 1 mmHg generates a shock wave travelling at ca. 6 times the speed of sound, at a pressure of ca. 50 mmHg and at a temperature of ca. 5000 K. Temperatures as high as 20 000 K have been produced in shock tubes. By this method a large number of pyrolysis and oxidation reactions have been studied.

Flames are examples of gaseous systems where a flow of gas combines with a self-supporting reaction to produce a steady-state situation. Special burners have been adapted so that spectroscopic monitoring of species within the flame is possible.

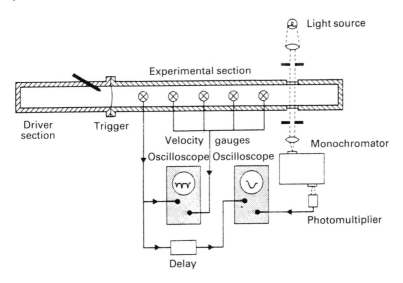

Fig. 2.2 — A typical layout of a conventional shock tube. Reproduced with permission from Bradley, *Fast Reactions* © 1975 Oxford University Press.

Obviously oxidation reactions are amenable to this approach. In particular, short-lived species in hydrocarbon oxidation, such as $C_2$, OH and $CHO^+$, have been identified in flames.

Plasmas are another example of high-energy gas phase systems. One type of plasma which has been used widely by chemists to generate reactive species is the flow discharge. These plasmas are normally excited by means of microwave, radio frequency or electric discharges, are of low pressure ($10^{-2}$ to 1 mmHg) and are at moderate temperatures (<ca. 300°C). They are initiated by free electrons (possibly, it has been suggested, from cosmic rays or background radiation) which cause electron-impact ionization of the gas. The plasma thus consists of electrons, ions (positive and negative) and neutral species (atoms, free radicals and molecules). Plasmas have been used to generate a wide range of free radicals, either directly or via the intermediacy of fluorine atoms, produced when $CF_4$ or $NF_3$ gas is flowed through a glow discharge. Further applications of plasmas are discussed in Chapters 8 and 9.

Intermediates of photochemical reactions have been much studied. Such reactions are discussed throughout this book; in particular, Chapter 3 focuses on the photochemistry of transition metal carbonyls, a field of research which has received a great deal of attention.

For molecules which are insensitive to ultraviolet–visible photolysis, photochemical reactions may be promoted by irradiation with vacuum ultraviolet light. As shown in Fig. 2.3, a system for 'windowless' photolysis has been developed, such a system being necessary to prevent absorption of vacuum ultraviolet light before it reaches the sample being irradiated. A typical source of vacuum ultraviolet light is a glow discharge of low-pressure argon or hydrogen, the latter having a strong emission at 121.6 nm. To initiate reactions at even higher energy, radiation such as

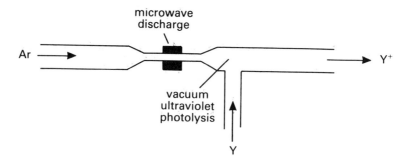

Fig. 2.3 — Schematic diagram of a 'windowless' resonance lamp for vacuum ultraviolet photolysis of a gaseous sample.

$\gamma$-rays or X-rays must be employed. In this way, for example, the radical $CH_3O$ has been produced by X-ray irradiation of solid methanol, $CH_3OH$ [9] (see section 8.2).

## 2.3   FLOW SYSTEMS

If an unstable species is generated continuously in a flow system, its decay establishes steady-state concentrations which decrease with distance downstream from the point of generation. At such points it is possible to detect spectroscopically the unstable species, alongside, of course, any stable products, unreacted reagents and carrier materials. Flow methods may be used to look at unstable species in the gas phase or in solution. They are ideally suited to kinetic measurements since, if the velocity of the flow of gas or liquid is known, it is possible to measure the rate of decay of the species under investigation.

The simplest type of flow system is the continuous flow method and a schematic diagram of such a system is given in Fig. 2.4. This method is ideal for looking at unstable intermediates, such as radicals and small molecules, formed in high-temperature vapours or by the action of a glow discharge. In the early years of such experiments the species in the flow were monitored by photographic recording of electronic absorption or emission spectra, giving only limited data by which the species could be characterized. Nowadays the use of tunable infrared lasers allows high-resolution vibrational spectra to be recorded of both small and quite large unstable molecules in the gas phase. Fig. 2.5 shows the high-resolution spectrum of the Q-branch of the bending vibration of the radical $FO_2$, produced in a flow tube reaction, and recorded using a tunable infrared diode laser [10].

Other means of detection of species in flow systems are available. Radicals, such as $CH_3^{\cdot}$ [11] or $CH_3O^{\cdot}$, [12], have been detected and characterized by flowing them, in a stream of carrier gas, through the cavity of an electron spin resonance (ESR) or a laser magnetic resonance (LMR) spectrometer. The latter approach is of particular importance in studying the far infrared spectra of unstable radicals. The deflection of a beam of molecules in an electric field may be measured, giving information about any permanent dipole moment of the species under investigation. Alternatively the beam may be flowed through a mass spectrometer. Mass spectrometric data provide information on the molecular weight and formula from observation of the parent ion

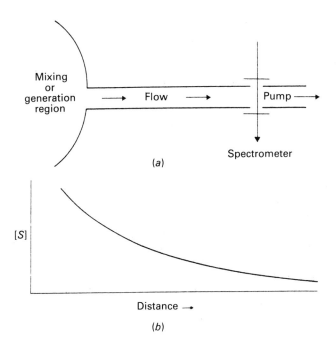

*(a)*

*(b)*

Fig. 2.4 — (a) Schematic diagram of a flow system; (b) change in concentration of a reaction intermediate $S$ with distance along the tube. Reproduced with permission from Ebsworth, Rankin and Cradock, *Structural Methods in Inorganic Chemistry* ©1987 Blackwell's Scientific Publications.

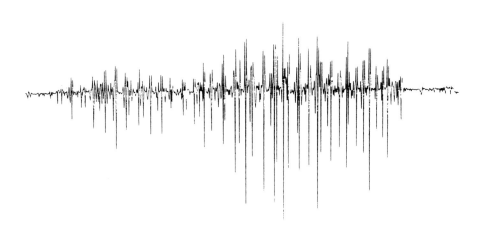

Fig. 2.5 — The Q-branch of the bending vibration of $FO_2$ prepared in a flow tube reaction. The band is centred near 578 cm⁻¹, and the whole spectrum shown covers only 1.5 cm⁻¹. Reproduced with permission from Yamada and Hirota, *J. Chem. Phys.* **80**, 4694 ©1984 American Institute of Physics.

peak, on molecular structure, from fragmentation patterns, and on thermodynamic parameters, from measurement of appearance potentials. Such measurements have been made on the molecule OPCl, formed by the reaction of $OPCl_3$ and silver at 1100 K [7]. Lastly the structure of high-temperature molecules in a flow of the vapour may be determined by gas phase electron diffraction. However, electron diffraction data from high-temperature molecules have often proved difficult to interpret, partly because of the high vibrational excitation of the molecules at these temperatures. Examples of such structure determinations are provided by the molecules $AlCl_3$ [13] and $FeCl_2$ [14] discussed in Chapter 7.

Before moving on from flow systems, it is pertinent to consider briefly the supersonic molecular beam. Here a flow of gas is expanded through a nozzle or orifice, into a vacuum, in such a way that the flow diverges and becomes supersonic. As the gas expands so the temperature falls, and molecules within the beam may be studied at very low temperatures. Section 4.10 discusses the use of supersonic molecular beams to study small metal clusters, and their reactions with gases such as hydrogen, oxygen and methane [15].

## 2.4 FLASH PHOTOLYSIS

If a short-lived species is suddenly generated in a non-flowing sample, it is possible to monitor the species spectroscopically some time after its generation, but before its decay is complete, provided that a very fast spectroscopic technique is available. This is the principle of flash photolysis, an invention of Norrish and Porter, which they first reported in 1949, and for which they shared a Nobel prize in 1967. The basic experimental arrangement is shown in Fig. 2.6. Generation of reactive intermediates is photochemical. In the original apparatus this was achieved by means of an ultraviolet lamp producing a very intense flash of light, typically of ca. 20 μs duration. The concentration of a reactive intermediate was monitored by means of a second

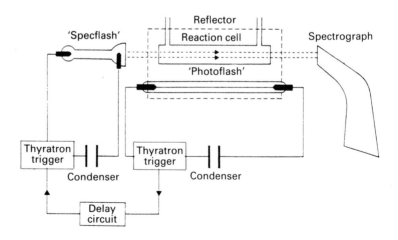

Fig. 2.6 — Diagram of a flash photolysis apparatus. Reproduced with permission from Bradley, *Fast Reactions* © 1975 Oxford University Press.

lamp (the so-called 'specflash') producing a longer less-intense flash of light at a wavelength which is absorbed by the species under investigation. Such a system is ideally suited to monitoring reaction kinetics, since the growth and decay of a particular reaction intermediate can be followed with time. The spectra recorded in this way, however, are of only limited use in identifying the reactive intermediates present in the system, since they give little structural information. Thus ambiguities have arisen in the interpretation of flash photolysis results (see section 3.3 and 5.4).

Modern flash photolysis experiments generally use lasers. One development, which is of particular interest to those concerned with reaction kinetics, is that of nanosecond flash photolysis. An arrangement for such an experiment is shown in Fig. 2.7. A pulse of light from the laser initiates the photolysis. However, some of the light from the laser is directed by means of the beam splitter, via a movable mirror to a vessel containing a fluorescent solution. Fluorescence from the solution gives a continuum of light in the visible region, and this is used as the spectroscopic source for monitoring the reaction intermediates. By adjusting the position of the movable mirror and delay between the 'photoflash' (of laser light) and the 'specflash' (of visible light from the fluorescent solution) can be altered.

Of great interest is the use of tunable infrared lasers to monitor the species produced in an ultraviolet flash photolysis experiment. In this way it is possible to

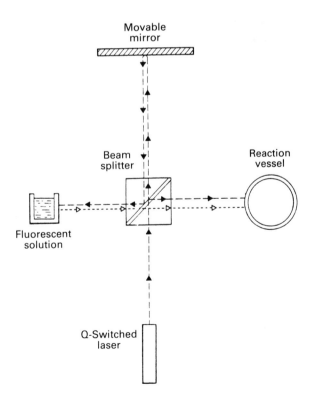

Fig. 2.7 — An arrangement for nanosecond flash photolysis. Reproduced with permission from Bradley, *Fast Reactions* © 1975 Oxford University Press.

study, not only the kinetics of a chemical reaction, but also to obtain infrared spectra of the intermediates involved. Thus considerable structural information about the intermediates is obtained. The experiment is not, however, trivial. The infrared laser is tuned to a particular frequency, and, following a photolytic flash from a pulsed ultraviolet source, kinetic measurements are made on the sample at that frequency. The monitoring infrared frequency is changed, and the experiment repeated. Data are thus accumulated over a particular region, but a large number of points are needed, even to cover a relatively small spectral region such as the $\nu(c{-}o)$ region. Elegant experiments of this type have been carried out by the group of Turner and Poliakoff at Nottingham [16], and the reader is referred to section 3.13 for a brief outline of some of their work.

## 2.5 MATRIX ISOLATION

A third means of recording spectra of short-lived species is to increase their lifetimes, by cooling the sample in an inert solid environment. The principle behind this approach lies in the fact that while the solid environment prevents biomolecular reactions (provided that the sample is at sufficiently high dilution), the low temperatures used greatly reduce the possibility of unimolecular decomposition. Typical solid host materials are glassy media, such as solid hydrocarbons or freons at liquid nitrogen temperature (77 K), or frozen gases (argon or nitrogen are often used) at lower temperatures (typically in the region 10–20 K).

Fig. 2.8 shows a standard arrangement for a matrix-isolation apparatus. The low temperatures are often achieved by the use of microrefrigerators using helium as the working fluid. Alternative coolants are liquid helium, or, less commonly, liquid

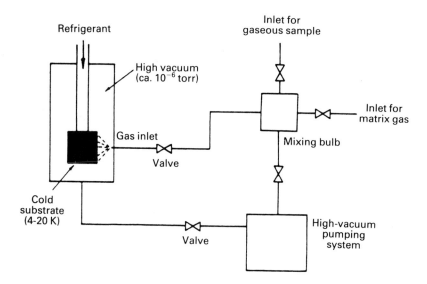

Fig. 2.8 — Schematic diagram of a matrix-isolation apparatus. Reproduced with permission from Almond and Downs, *Advances in Spectroscopy*, **17**, 1 ©1989 John Wiley.

hydrogen. The cold substrate, the nature of which depends on the type of spectroscopy being used to monitor the sample (see Table 2.1), is enclosed in a shroud which
is maintained at high vacuum.


**Table 2.1** — Various substrates employed for spectroscopic detection of matrix-
isolated species

| Spectroscopic method | Substrate |
|---|---|
| Ultraviolet absorption or emission; magnetic circular dichroism | Quartz or lithium fluoride window |
| Infrared transmission | Caesium iodide or other alkali halide window |
| Infrared reflectance | Polished metal plate, e.g. Cu |
| Raman | Polished metal plate, e.g. Cu, brass, stainless steel or Cs1 window |
| ESR | Sapphire rod |
| NMR | Sapphire rod |
| Mössbauer | Be disc |
| EXAFS | Al foil |
| SIMS | Cu plate |


Preparation of matrix-isolated samples generally follows one of two routes. First,
the short-lived species may be formed in the gas phase by photolysis, discharge,
pyrolysis, etc., then condensed with a large excess (typically a thousand-fold) of the
matrix material. In the second approach a stable precursor is trapped within the
matrix, and the unstable species are generated *in situ* by photolysis, radiolysis, or by
chemical reaction often induced by warming the matrix slightly to allow species
within the matrix to migrate more easily. These methods are illustrated schematically
in Fig. 2.9, while Fig. 2.10 shows sections through the vacuum shroud of a matrix-
isolation apparatus, as set up for infrared detection of photolytically generated
species.

Once trapped within a matrix the lifetimes of many species, which exist only
transiently under normal conditions, are so extended that the species can be studied
at leisure by conventional spectroscopy. Infrared spectroscopy has been the most
widely favoured method. It is always desirable to complement infrared spectra by
Raman measurements, and Raman spectroscopy has been applied to a wide range of
matrix-isolated molecules — although there are problems of photosensitivity when
exciting Raman scattering from a matrix sample. Resonance Raman spectra have
been excited from several matrix-isolated samples, as has laser-induced fluorescence, while UV-visible and ESR measurements have also been widely used,
although the latter technique is necessarily limited to paramagnetic species. In recent
years the repertoire of physical methods has been increased to include magnetic
circular dichroism (MCD), Mössbauer spectroscopy, EXAFS and even NMR and

photoelectron spectroscopy. Infrared spectroscopy will, however, remain the major tool for studying matrix-isolated species, particularly as relatively cheap and compact Fourier transform instruments are now available.

Throughout this book, many examples of the use of matrix isolation are given, but these represent only a small fraction of the chemical systems which have been studied by the technique. Of the various methods available, matrix isolation is probably the best for studying the structure of short-lived molecules. But it does have its drawbacks. First, it gives no information at all about the kinetics of a chemical process. Second, the question must always be asked: to what extent are species observed in a low-temperature matrix analogous to transient intermediates in more normal chemical environments? Although it appears that most molecules are not much perturbed on moving from gas to matrix-isolated phases, this second point must always remain a worry, if matrix-isolation results are taken in 'isolation' without recourse to any other technique.

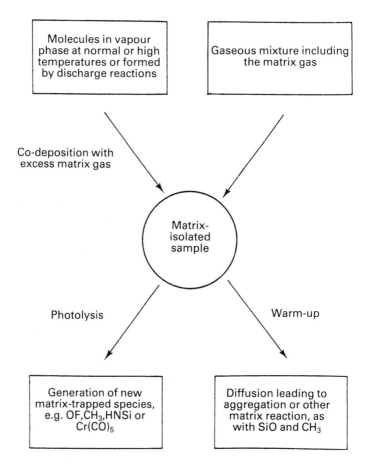

Fig. 2.9 — Preparation and treatment of a matrix-isolated sample. Reproduced with permission from Almond and Downs, *Advances in Spectroscopy*, **17**, 1 ©1989 John Wiley.

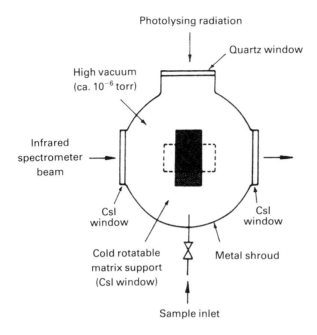

Fig. 2.10 — Section through the vacuum shroud of a matrix-isolation apparatus suitable for infrared measurements. Reproduced with permission from Almond and Downs, *Advances in Spectroscopy* **17**, 1 ©1989 John Wiley.

## 2.6   STUDIES IN LOW-TEMPERATURE SOLUTION

One way of answering these doubts about matrix-isolation experiments is to investigate the infrared spectra of short-lived molecules, not within a solid matrix, but in a low-temperature liquid phase solution. In this way the reactive species are 'slowed down' rather than 'stopped'. Thus, while conventional spectroscopic monitoring is possible, kinetic information about the species under investigation is also obtained. In this way, a 'bridge' is formed between, on the one hand, room temperature flash photolysis work, and, on the other, low-temperature matrix-isolation spectroscopy.

A typical solvent is liquid xenon. An advantage of this solvent is that it does not have any infrared absorptions, so long path-length cells (up to about 25 mm) may be employed, allowing weak absorptions of the transient intermediates to be detected. This method also allows the thermal stability of short-lived molecules to be studied. The liquid xenon solution may be warmed, the limitation being the capacity of the spectroscopic cell to withstand high pressure. Using specially designed cells it is now possible to look at infrared spectra in liquid xenon solution at room temperature. Some experiments using liquid xenon as a solvent are discussed in Section 3.12.

## 2.7   CHEMICAL TRAPPING

To end this chapter it is pertinent to consider an approach which is not a means of directly observing a short-lived species, but rather is a means of inferring the presence of the species. This is the technique of chemical trapping. Here a reagent is added to the reaction mixture which forms a stable product with the reactive intermediate, which is particularly characteristic of that intermediate. The stable product may then be characterized by standard chemical and spectroscopic analysis. In this way the nature of the intermediate may be determined. Some examples of the use of this approach in studying divalent silicon species are discussed in Chapter 5. This approach does, of course, suffer from ambiguity in the interpretation of results, unless a range of different stable products can be produced from the same intermediate with a number of different trapping reagents.

## 2.8 SUMMARY

The study of reaction mechanisms and of short-lived molecules is an ever-expanding area of chemistry. The output of just one technique discussed in this chapter — that of matrix isolation — is continually increasing. Recent literature searches [17] have shown that while about 900 papers including matrix-isolation results were published between 1954 and 1976, some 2500 papers on the same topic appeared in the ten years from 1977. This increasing activity is mirrored by the other experimental methods mentioned in this chapter. Thus much information about short-lived molecules is now available, and it is certain that our knowledge about such species will continue to increase in the future.

What is becoming clear is that the best and most unambiguous studies of transient species encompass a variety of experimental methods. Thus spectroscopic characterization of the species is required, both to establish its identity and to obtain structural information. This characterization is perhaps best carried out by the matrix-isolation technique, where a variety of conventional spectroscopic methods can be brought to bear on the species under investigation. However, matrix isolation gives no kinetic information, and its is impossible to understand properly a reaction mechanism without knowledge of the kinetics of the reaction, even if reaction intermediates are well characterized. Kinetic information is obtained from flash photolysis and flow experiments. To bring together the kinetic and spectroscopic data from room temperature solution and low-temperature matrix studies, the use of low-temperature solutions is important.

To take one example of transition metal carbonyl photochemistry, the laboratory of Turner and Poliakoff at the University of Nottingham is, in this respect, particularly well equipped. Here solid matrices, low-temperature solutions and time-resolved infrared spectra of room temperature solutions can all be studied. The combination of the three techniques has allowed some quite complex photochemical mechanisms to be explored. A discussion of some of this work will be found in Chapter 3.

## REFERENCES

[1] R. Criegee and G. Wenner, *Liebigs Ann. Chem.* (1949), **564**, 9; R. Criegee, *Angew. Chem. Int. Ed. Engl.* (1975), **14**, 745.

[2] G. A. Bell and I. R. Dunkin, *J. Chem. Soc., Chem. Commun.* (1983), 1213; O. L. Chapman and T. C. Hess, *J. Am. Chem. Soc.* (1984), **106**, 1842; I. R. Dunkin and C. J. Shields, *J. Chem. Soc., Chem. Commun.* (1986), 154.

[3]  I. R. Beattie and J. E. Parkinson, *J. Chem. Soc., Dalton Trans.* (1983), 1185.

[4]  S. A. Arthers, I. R. Beattie, R. A. Gomme, P. J. Jones and J. S. Ogden, *J. Chem. Soc., Dalton Trans.* (1983), 1461; I. R. Beattie and J. E. Parkinson, *J. Chem. Soc., Dalton Trans.* (1984), 1363.

[5]  J. S. Anderson and J. S. Ogden, *J. Chem. Phys.* (1969), **51**, 4189.

[6]  J. S. Ogden and S. J. Williams, *J. Chem. Soc., Dalton Trans.* (1982), 825.

[7]  M. Binnewies, M. Lakenbrink and H. Schnöckel, *Z. Anorg. Allg. Chem.* (1983), **497**, 7; H. Schnöckel and S. Schunck, *Z. Anorg. Allg. Chem.* (1987), **548**, 161.

[8]  D. W. Green and K. M. Ervin, *J. Mol. Spectrosc.* (1981), **89**, 145.

[9]  M. Iwasaki and K. Toriyama, *J. Am. Chem. Soc.* (1978), **100**, 1964.

[10] C. Yamada and E. Hirota, *J. Chem. Phys.* (1984), **80**, 4694.

[11] R. W. Fessenden and R. H. Schuler, *J. Chem. Phys.* (1963), **39**, 2147.

[12] H. E. Radford and D. K. Russell, *J. Chem. Phys.* (1977), **66**, 2222.

[13] M. Hargittai and I. Hargittai, *J. Mol. Spectrosc.* (1984), **108**, 155.

[14] I. Hargittai, J. Tremmel and G. Schultz, *J. Mol. Struct.* (1975), **26**, 116; E. Vajda, J. Tremmel and I. Hargittai, *J. Mol. Struct.* (1978), **44**, 101.

[15] M. Geusic, M. D. Morse and R. E. Smalley, *J. Chem. Phys.* (1985), **82**, 590; S. C. Richtsmeier, E. K. Parks, K. Liu, L. G. Pobo and S. J. Riley, *J. Chem. Phys.* (1985), **82**, 3659; R. L. Whetten, D. M. Cox, D. J. Trevor and A. Kaldor, *J. Phys. Chem.* (1985), **89**, 566.

[16] B. D. Moore, M. Poliakoff, M. B. Simpson and J. J. Turner, *J. Phys. Chem.* (1985), **89**, 850; A. J. Dixon, S. J. Gravelle, L. J. van de Burgt, M. Poliakoff, J. J. Turner and E. Weitz, *J. Chem. Soc., Chem. Commun.* (1987), 1023.

[17] D. W. Ball, Z. H. Kafafi, L. Fredin, R. H. Hauge and J. L. Margrave, *A Bibliography of Matrix Isolation Spectroscopy 1954–1985;* Rice University Press: Houston, Texas (1988); M. J. Almond and A. J. Downs, *Spectroscopy of Matrix-isolated species*, vol. 17 of *Advances of Spectroscopy*, R. J. H. Clark and R. E. Hester (eds); John Wiley: Chichester (1989).

# 3

# Photochemistry of metal carbonyls

## 3.1 INTRODUCTION

Transition metal carbonyls are excellent molecules for photochemical study. They are mostly readily soluble in inert solvents, have ultraviolet absorption with high extinction coefficients and have high quantum yields for photochemical reactions. In particular they are highly susceptible to photochemical substitution of carbonyl groups. In general, moreover, such molecules are strong infrared absorbers and strong in Raman scattering, especially in the region 1600–2200 cm$^{-1}$ associated with CO stretching fundamentals. Thus vibrational spectroscopy can readily be applied to monitoring reactions of these molecules.

## 3.2 SOLUTION STUDIES

A classical example of a photochemical substitution reaction of a metal carbonyl occurs when chromium hexacarbonyl, $Cr(CO)_6$, is subjected to photolysis in cyclo-hexane solution containing triphenylphosphine. In this reaction, the molecule $Cr(CO)_5(PPh_3)$ is produced with the simultaneous evolution of CO.

$$Cr(CO)_6 + PPh_3 \xrightarrow[\text{cyclohexane}]{h\nu} Cr(CO)_5(PPh_3) + CO \uparrow$$

A second well-known example is the deposition of the solid binuclear metal carbonyl $Fe_2(CO)_9$ when $Fe(CO)_5$ is irradiated in glacial acetic acid solution.

$$Fe(CO)_5 \xrightarrow[\text{glacial acetic acid}]{h\nu} Fe_2(CO)_9 \downarrow$$

In the early 1960s, Sheline and his co-workers at Florida State University carried out experiments designed to study the replacement of the carbonyl groups of metal carbonyls by electron donor molecules. It was found, for example, that while a solution of tungsten hexacarbonyl, $W(CO)_6$, in $n$-hexane, containing acetonitrile,

CH$_3$CN, was colourless, exposure of this solution to ultraviolet light resulted in its turning bright yellow almost immediately [1]. At the same time new carbonyl stretching bands appeared in the 1700–2200 cm$^{-1}$ region of the infrared spectrum of the solution. After 20 s exposure to ultraviolet light the principal product observed was W(CO)$_5$(CH$_3$CN) together with a smaller amount of W(CO)$_4$(CH$_3$CN)$_2$ and a low yield of W(CO)$_3$(CH$_3$CN)$_3$. However, the yield of W(CO)$_3$(CH$_3$CN)$_3$ could be increased considerably if the solution, after being allowed to stand for 16 min, was re-exposed to ultraviolet light for 2.5 min. Interestingly, if the solution were finally left to stand for 3 min, the concentration of W(CO)$_6$ *increased*. The products W(CO)$_5$(CH$_3$CN), W(CO)$_4$(CH$_3$CN)$_2$ and W(CO)$_3$(CH$_3$CN)$_3$ are all stable, and could be isolated, allowing unequivocal characterization to be made. These spectral changes are listed in Table 3.1, and are illustrated in Fig. 3.1, while the reactions can be summarized as follows:

$$W(CO)_6 + CH_3CN \rightleftharpoons W(CO)_5(CH_3CN) + CO$$
$$\big\updownarrow CH_3CN$$
$$W(CO)_3(CH_3CN)_3 + CO \underset{CH_3CN}{\rightleftharpoons} W(CO)_4(CH_3CN)_2 + CO$$

The observation that W(CO)$_6$ was regenerated when the photolysed solutions were allowed to stand, led Sheline and his co-workers to conclude that the 16-electron neutral metal carbonyl, W(CO)$_5$, was involved in the reaction. This intermediate can react with the electron-pair donor molecule CH$_3$CN to yield the 18-electron product W(CO)$_5$CH$_3$CN, but can also combine with traces of CO trapped in the spectroscopic cell to yield the starting material W(CO)$_6$. However, no direct spectroscopic evidence could be obtained for this short-lived intermediate.

Shortly afterwards these same workers extended their studies to the irradiation of W(CO)$_6$ in *n*-hexane solution containing various alkenes and alkynes [2]. A similar

**Table 3.1** — Bands observed in 1700–2200 cm$^{-1}$ region of the infrared spectrum on irradiation of W(CO)$_6$ in *n*-hexane solution containing acetonitrile

| $\nu$/cm$^{-1}$ | Assignment |
|---|---|
| 2140 | CO |
| 2075 | W(CO)$_5$(CH$_3$CN) |
| 2025 | W(CO)$_4$(CH$_3$CN)$_2$ |
| 1980 | W(CO)$_6$ |
| 1940 | W(CO)$_5$(CH$_3$CN) |
| 1910 | W(CO)$_3$(CH$_3$CN)$_3$ |
| 1897 | W(CO)$_4$(CH$_3$CN)$_2$ |
| 1835 | W(CO)$_4$(CH$_3$CN)$_2$ |
| 1790 | W(CO)$_3$(CH$_3$CN)$_3$ |

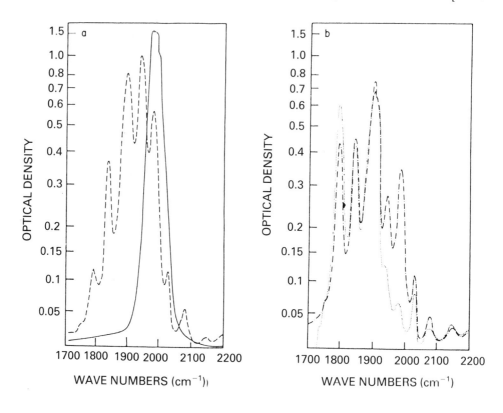

Fig. 3.1 — Infrared and ultraviolet-visible spectra of W(CO)₆ in CH₃CN solution. (a) ———
before irradiation; – – – – after 20 s broad-band photolysis; (b) ········ after standing for 16 min
followed by 2.5 min broad-band photolysis; —·—·—· after standing for a further 3 min.
Reproduced with permission from Dobson *et al.*, *Inorg. Chem.*, **1**, 526 © 1962 American
Chemical Society.

picture emerged. For example, photolysis of $W(CO)_6$ in the presence of ethylene or
acetylene yields the molecules $W(CO)_5(C_2H_4)$ and $W(CO)_5(C_2H_2)$, respectively.
The carbonyl stretching frequencies of these complexes are given in Table 3.2,
together with the assignments of the bands.

**Table 3.2** — Infrared absorptions of $W(CO)_5(C_2H_4)$ and
$W(CO)_5(C_2H_2)$ in *n*-hexane solution

| Complex | $v/cm^{-1}$ | | |
|---|---|---|---|
| | $a_1$ | e | $a_1$ |
| $W(CO)_5(C_2H_4)$ | 2088(w), | 1953(s) | 1973(ms) |
| $W(CO)_5(C_2H_2)$ | 2095(w), | 1967(s) | 1952(ms) |

It is interesting to note the relative positions of the more intense of the two $a_1$ bands for each complex. This absorption is produced by stretching of the unique carbonyl group *trans* to the substituent, in the complex **1**.

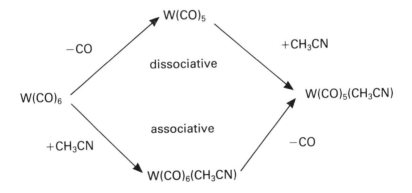

$$L = C_2H_4 \text{ or } C_2H_2$$

**(1)**

The higher frequency of this vibration in the molecule $W(CO)_5(C_2H_4)$ as compared to that in $W(CO)_5(C_2H_2)$ reflects the fact that acetylene is a better charge donor and a weaker charge acceptor than ethylene. Thus, in $W(CO)_5(C_2H_2)$ more charge is donated to the $\pi^*$ orbital of the *trans* CO group and the frequency of the $a_1$ vibration associated with this group is lowered.

Various mechanistic interpretations can be proposed for the results of these, and other similar studies. In the first case an associative or a dissociative mechanism could, in principle, be operating.

Secondly, if a dissociative mechanism is preferred, it is conceivable that this could proceed via a radical or an ionic mechanism:

$$W(CO)_6 \xrightarrow{h\nu} W(CO)_5 + CO$$

$$W(CO)_6 \xrightarrow{h\nu} W(CO)_5^- + CO^+$$

The observations of Sheline *et al.* [1] that $W(CO)_6$ could be regenerated in some of their photolysis experiments was circumstantial evidence in favour of the radical dissociative process. However, if this mechanism is correct, it is important to establish the presence of the pentacarbonyl intermediate. The most obvious way to tackle this problem is by conventional flash photolysis experiments in solution.

### 3.3  FLASH PHOTOLYSIS STUDIES

It was shown that when a cyclohexane solution of $Cr(CO)_6$ (approximately $10^{-4}$ mol $dm^{-3}$) was subjected to flash photolysis, two consecutive transient species were observed [3]. The first of these, **2**, has an absorption maximum in the visible spectrum at 483 nm; it does not react with CO, but instead forms, with a half-life of 6 ms, the second intermediate, **3**, which has an absorption maximum of 445 nm and slowly recombines with CO, with a half-life of 25 s, to yield $Cr(CO)_6$. The original interpretation of these results was that **2** and **3** were isomers of the $Cr(CO)_5$ molecule. It was suggested that **2** was a square-pyramidal, $C_{4v}$, form of $Cr(CO)_5$, while **3** was a trigonal bipyramidal, $D_{3h}$, form of the same molecule.

However, the presence of a $D_{3h}$ form of $Cr(CO)_5$ was questioned, and later flash photolysis experiments were performed using an apparatus with a higher time resolution (flash half-width $<3 \mu s$) and with more extensively purified solvent [4]. Under these conditions a new intermediate, **4**, was observed, which shows a broad band in the visible absorption spectrum with maximum absorption at 503 nm. It has a lifetime of $>200 \mu s$ and reacts to form a second species, whose visible absorption spectrum ($\lambda_{max} = 445$ nm) is similar to that previously observed for **3**. Like **3** this second intermediate decays with a lifetime greater than 1 s to reform the hexacarbonyl. When less rigorously purified solvent was used, **4** also yielded a third intermediate, whose visible absorption spectrum ($\lambda_{max} = 470$–480 nm) was similar to that observed in the earlier experiments. The interpretation of these results is that **4** is $Cr(CO)_5$, while both **2** and **3** are more long-lived intermediates which arise from reaction of $Cr(CO)_5$ with impurities in the solvent. The reactions can be summarized as follows, where X and Y are impurities.

$$Cr(CO)_6 \longrightarrow Cr(CO)_5 + CO$$
(4)

$$Cr(CO)_5 + CO \longrightarrow Cr(CO)_6$$
$$Cr(CO)_5 + X \longrightarrow Cr(CO)_5 X$$
(2)

$$Cr(CO)_5 + Y \longrightarrow Cr(CO)_5 Y$$
(3)

These results indicate the difficulties encountered in interpreting the results of flash photolysis experiments, involving UV–visible detection, for a complex system, such as this. First, no structural information can be gained from such studies. Thus, while **4** is likely to be $Cr(CO)_5$, the structure adopted by this species cannot be ascertained. Nor is it possible to say whether a cyclohexane solvent molecule is loosely co-ordinated to the metal centre in **4**. Furthermore, minute traces of impurity in the solvent interfere strongly with the spectral behaviour. Much more structural information could be obtained from *infrared* spectra. The best way to record infrared spectra of transient intermediates is by using the matrix-isolation technique to trap the intermediate. With this aim in mind, experiments were carried out in which it was attempted to generate $Cr(CO)_5$ in rigid glasses.

## 3.4 EXPERIMENTS IN RIGID GLASSES

Preliminary, encouraging results on the photolysis of $Cr(CO)_6$, $Mo(CO)_6$ and $W(CO)_6$ in $1:4$ isopentane–methylcyclohexane glasses at $-180°C$ were obtained by the group of Sheline [5]. These workers found that ultraviolet irradiation of $M(CO)_6$ (M = Mo or Cr) molecules, under these conditions, yields in each case a molecule which shows three infrared absorptions in the $\nu_{C-O}$ region. The number of these bands, and their relative intensities, is consistent with the interpretation that the intermediates are $M(CO)_5$ molecules with $C_{4v}$ symmetry. The frequencies of these bands are listed in Table 3.3. These workers also reported that a new species was produced when $Fe(CO)_5$ was subjected to photolysis under similar conditions.

**Table 3.3** — Infrared absorptions of $Cr(CO)_5$ and $W(CO)_5$ trapped in $1:4$ isopentane–methylcyclohexane glass at $-180°C$

| Species | $v/cm^{-1}$ |
|---------|-------------|
| $Cr(CO)_5$ | 2088(w), 1955(s), 1928(m) |
| $Mo(CO)_5$ | 2093(w), 1960(s), 1922(m) |
| $W(CO)_5$ | 2092(w), 1952(s), 1924(m) |

More recently, the scope of photochemical studies on metal carbonyl complexes in hydrocarbon glasses has been considerably extended by the work, among others, of Braterman, at the University of Glasgow. For example, he has shown that photolysis at wavelengths greater than 305 nm of the complex $Mo(CO)_5PCx_3$ (Cx = cyclohexyl) in $4:1$ isopentane–methylcyclohexane glass at 77 K forms two new species [6]. The first of these, **5**, shows CO stretching bands at 2027.5 (m), 1916 (sh), 1907 (s) and 1864 (m) $cm^{-1}$, while the second species, **6**, shows a single CO stretching band at 1883 $cm^{-1}$. While irradiation of $Mo(CO)_5PCx_3$ at wavelengths above 420 nm has no effect, it was found that **6** is readily converted to **5** by photolysis at these wavelengths. The conclusion reached by these workers was that **5** and **6** are isomers of the molecule $Mo(CO)_4(PCx_3)$. The four-band pattern observed for **5** is expected for a species of the type *cis*-$M(CO)_4LL'$, while in **6** the 1883 $cm^{-1}$ band is likely to arise from the intense e mode of a complex of the type *trans*-$M(CO)_4LL'$.

The following structures can, therefore, be assigned to the species **5** and **6**.

(□ = vacancy, or co-ordinated solvent molecule)

This example illustrates only one of many such studies on unstable metal carbonyl molecules trapped in rigid glasses. The infrared spectra obtained from these studies have been of great importance in identifying intermediates in these reactions, and in determining their structures. Two problems, however, remain. The first is to what extent can such glasses be considered as being chemically inert? A co-ordinated solvent molecule is likely to perturb, for example, the vibrations of a molecule such as $Cr(CO)_5$. Secondly the problem of rigorous purification of solvents arises. Even slight traces of impurity are likely to affect substantially the spectra of the species being investigated. The most obvious way to overcome these difficulties is to move to an inert gas matrix at lower temperatures.

### 3.5 MATRIX-ISOLATION STUDIES ON $M(CO)_5$ (M = Cr, Mo OR W) MOLECULES

In 1971 Turner and Poliakoff published the results of experiments in which they had succeeded in producing the 16-electron $M(CO)_5$ molecules (M = Cr, Mo or W) by photolysis of $M(CO)_6$ in argon or methane matrices at 20 K [7]. Fig. 3.2 illustrates the spectra obtained from an experiment involving ultraviolet photolysis of $Cr(CO)_6$ in a methane matrix. $Cr(CO)_6$ has octahedral symmetry, and, as such, shows only one infrared active band, of $t_{1u}$ symmetry, in the C–O stretching region of the spectrum. The slight splitting of this band, seen in Fig. 3.2, is caused by matrix-site effects, that is the possibility that $Cr(CO)_6$ can occupy slightly different sites within the matrix lattice. Broad-band ultraviolet-visible photolysis caused decay in the intensity of this band. At the same time four new bands appear in this region of the spectrum. One of these (at $2140$ cm$^{-1}$) can be assigned to the vibration of uncoordinated molecular carbon monoxide. The other three can be assigned to the three vibrations ($2a_1 + e$) expected for a $Cr(CO)_5$ fragment of $C_{4v}$ symmetry. In a later study [8], Perutz and Turner carried out a detailed analysis of the spectrum of $Cr(CO)_5$, produced on photolysis of randomly substituted $Cr(^{13}CO)_x(^{12}CO)_{6-x}$, enriched with ca. 50% $^{13}CO$. These spectra were inconsistent with a $D_{3h}$ structure (3)

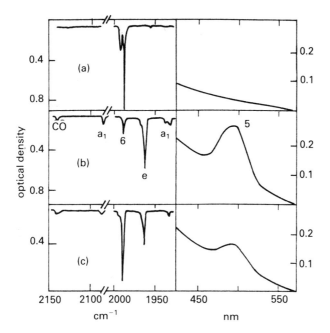

Fig. 3.2 — Infrared and ultraviolet-visible spectra of Cr(CO)₆ in a CH₄ matrix at 20 K: (a) after sample deposition; (b) after 30 s broad-band photolysis; (c) after 1.5 min photolysis at λ>330 nm. Bands marked '5' are due to Cr(CO)₅, and those marked '6' are due to Cr(CO)₆. Reproduced with permission from Turner *et al.*, *Pure Appl. Chem.*, **49**, 271 © 1977 Pergamon.

for the pentacarbonyl fragment, but could be fitted accurately using a $C_{4v}$ structure (**2**). From intensity data, axial–radial bond angles of 90–95° were calculated.

```
        CO
 90–95°  |
OC ~⌐ ↗  ⌐ - CO
       Cr
OC ◣      ◣ CO
      (2)
```

```
        CO
         |
OC ──── Cr ⁻ CO
         |
        CO
        ◣ CO
      (3)
```

The frequencies of the infrared absorptions of the C–O stretching modes of Cr(CO)₅ in a mixed 1:4 isopentane–methylcyclohexane glass at 96 K [5] and in a methane matrix at 20 K [7] are compared in Table 3.4. It can be seen that the values are in quite good agreement.

Alongside these infrared absorptions, Cr(CO)₅, when isolated in a methane matrix, at 20 K, exhibits a visible absorption band at 489 nm. It was found that while the vibrational spectrum of the Cr(CO)₅ fragment varies little on changing from one matrix to another, the position of the visible band of Cr(CO)₅ is extremely sensitive to the matrix material, varying from 624 nm in neon to 489 nm in methane and

**Table 3.4** — Infrared absorptions of $Cr(CO)_5$ trapped
in a 1:4 isopentane–methylcyclohexane glass at 96 K
and in a methane matrix at 20 K

| Mode | $v/cm^{-1}$ (glass) | $v/cm^{-1}$ (matrix) |
|------|------|------|
| $a_1$ | 2088 | 2088 |
| e | 1955 | 1961 |
| $a_1$ | 1928 | 1932 |

487 nm in xenon. Intermediate values of 545 and 560 nm are seen for argon and sulphur hexafluoride matrices, respectively. The question arose as to whether this variation results from a general solvent effect, or from a specific interaction of a matrix molecule into the vacant co-ordination site of $Cr(CO)_5$.

Fig. 3.3 illustrates the results of an experiment carried out in a mixed matrix of neon containing 2% xenon. The most striking feature of these spectra is that two distinct absorptions are seen. If the shift in frequency on moving from neon to xenon matrices had been the result of simply a general solvent effect, it would be expected that a mixed neon/xenon matrix would give a single absorption located somewhere between the positions of the absorptions in the pure matrices. The observation of two distinct bands in the mixed matrix demonstrates that a specific interaction takes place

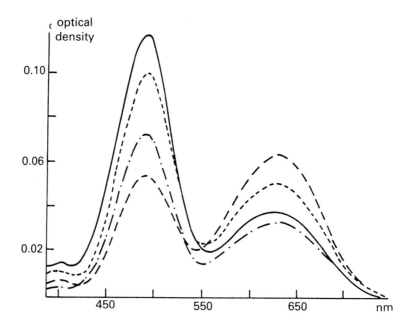

Fig. 3.3 — Visible spectra of $Cr(CO)_5$ in a Ne/2% Xe matrix at 4 K: ——— after deposition of $Cr(CO)_6$, then 15 s broad-band photolysis; – – – – after 5 min photolysis at $\lambda = 432$ nm; — — — after 40 min photolysis at $\lambda = 432$ nm; —·—·— after 7 min photolysis at $\lambda = 618$ nm. Reproduced with permission from Turner *et al.*, *Pure Appl. Chem.*, **49**, 271 © 1977 Pergamon.

with both neon and xenon, a point which is further emphasized by the photochemical behaviour of the two complexes. From the spectra shown in Fig. 3.3 it is clear that a reversible substitution reaction is taking place.

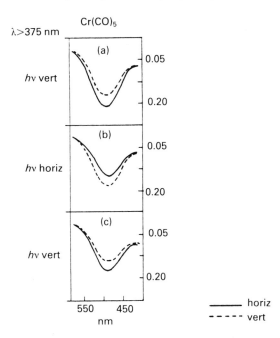

Other reactions may be promoted by irradiation into the visible band of an $M(CO)_5$ fragment (M = Cr, Mo or W), which has been generated by ultraviolet photolysis of $M(CO)_6$ in a low-temperature matrix. One example is the uptake of a photoejected CO molecule from the matrix to regenerate the parent hexacarbonyl.

$$M(CO)_5 + CO \xrightarrow[20\ K]{hv\ (visible)} M(CO)_6$$

The mechanism of this reaction has been the subject of some elegant experiments involving polarized light both for photolysis and for spectroscopic measurements [9]. The visible absorption spectra obtained from these experiments are illustrated in Fig. 3.4. Unpolarized ultraviolet photolysis of $Cr(CO)_6$ in pure methane matrices

Fig. 3.4 — Results of irradiating matrix-isolated $Cr(CO)_5$ with polarized light at $\lambda = 375$ nm. Reproduced with permission from Turner *et al.*, *Pure Appl. Chem.*, **49**, 271 © 1977 Pergamon.

yields randomly oriented $CH_4 \ldots Cr(CO)_5$ molecules. These molecules can then be subjected to polarized visible ($\lambda > 375$ nm) photolysis, and from the spectra illustrated in Fig. 3.4, it can be seen that linear dichroism develops in the visible absorption band. Rotating the irradiating beam through $90°$ ($hv$ vert. $\rightarrow hv$ horiz.) reverses the spectral dichroism, but this can be returned to the original value on turning the polarized radiation back again. The absorbances, measured in these experiments, showed that each reversal was accompanied by an *increase* in optical density of the polarization relative to the previous spectrum. This shows that the $Cr(CO)_5$ molecules are undergoing photo-orientation; that is, irradiation with polarized light causes the molecules to reorient to configurations where they have a lower probability of absorbing the incident radiation. This can be summarized by the following reaction.

$$CH_4 \ldots Cr(CO)_5 \xrightarrow[\text{horiz.}]{hv} CH_4 \ldots Cr \xrightarrow[\text{vert.}]{hv} Cr$$

CH$_4$

These experiments show that the matrix holds the $Cr(CO)_5$ fragment rigid at the beginning and end of the photolysis period, but that mobility is possible during irradiation. The most likely explanation is that visible photolysis causes the $Cr(CO)_5$ fragment to be excited from the ground state, $C_{4v}$ configuration, to a $D_{3h}$ excited state. The excited state can then relax back to a $C_{4v}$ geometry, but reorientation of the molecule is possible.

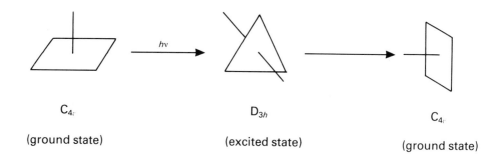

| $C_{4v}$ | $D_{3h}$ | $C_{4v}$ |
|:---:|:---:|:---:|
| (ground state) | (excited state) | (ground state) |

It is, of course, now possible to propose a mechanism by which a $Cr(CO)_5$ fragment might take up a free CO molecule, on visible photolysis, to form $Cr(CO)_6$. This is via photoreorientation causing the vacant site of $Cr(CO)_5$ to 'move round' the molecule, such that it may be directed towards any adjacent CO molecule, trapped in the matrix cage. In this way a chemical bond will be formed between CO and the $Cr(CO)_5$ fragment. Alternatively, if the vacant site of the final product is directed towards a matrix methane molecule then $CH_4 \ldots Cr(CO)_5$ will be formed. This 'product' is, however, still labile to visible photolysis, so reorientation may continue.

Thus a reaction is possible with any CO molecule which is adjacent to the $Cr(CO)_5$ fragment, so the recombination reaction:

$$Cr(CO)_5 + CO \longrightarrow Cr(CO)_6$$

is promoted by visible photolysis.

## 3.6   MATRIX-ISOLATION STUDIES ON Fe(CO)₄

Ultraviolet photolysis of matrix-isolated $Fe(CO)_5$ yields the 16-electron fragment, $Fe(CO)_4$ [10]. Isotopic studies involving $^{13}CO$ have confirmed that this molecule has a $C_{2v}$ geometry, **7**.

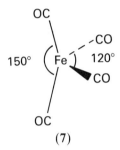

(7)

Bond angles have been calculated from infrared-absorption intensity measurements and are found to be 150° and 120° for $Fe(CO)_4$ trapped in a methane matrix. These values, which vary only slightly from matrix to matrix, show that $Fe(CO)_4$ is substantially distorted from tetrahedral geometry. They are close to the bond angles predicted by Burdett (135° and 110°) for the minimum energy geometry of $Fe(CO)_4$ in a *triplet* ground state [11]. $Fe(CO)_4$ is a $d^8$ system and in a tetrahedral geometry it would necessarily have a triplet ground state, this configuration would, however, be Jahn–Teller unstable, so would distort to a minimum-energy $C_{2v}$ configuration. For singlet $Fe(CO)_4$ the distortion would continue to square planar geometry, but further distortion of the $C_{2v}$ geometry of triplet $Fe(CO)_4$ is prevented by the rapid rise in energy of the highest energy $d$ orbital (see Fig. 3.5). Since $Fe(CO)_4$ has $C_{2v}$ geometry it is predicted to be paramagnetic. Studies involving the technique of magnetic circular dichroism (MCD) are consistent with a paramagnetic ground state for $Fe(CO)_4$, since the molecule shows a temperature-dependent MCD spectrum, whereas diamagnetic compounds do not show any temperature dependence of their MCD signals. $Fe(CO)_4$ is therefore the first binary transition metal carbonyl shown to have a 'high-spin' electronic ground state. The energy levels of the $d$ orbitals in $Fe(CO)_4$ in different geometries are shown in Fig. 3.5.

Near-infrared photolysis of matrix-isolated $Fe(CO)_4$ promotes a reaction with photoejected CO to regenerate $Fe(CO)_5$. For example,

$$Fe(CO)_4 + CO \xrightarrow[\text{near i.r.}]{h\nu} Fe(CO)_5$$

$Fe(CO)_4$

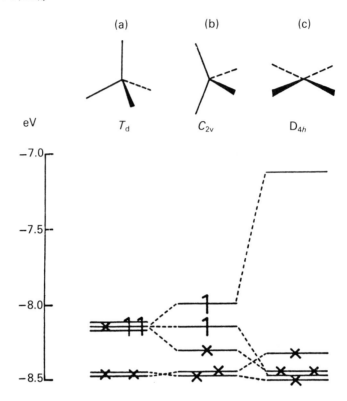

Fig. 3.5 — Ligand-field splitting diagram for the $d^8$ complex Fe(CO)$_4$ with $T_d$, $C_{2v}$ and $D_{4h}$ symmetries. Energy levels of the $d$ orbitals of Fe(CO)$_4$ in tetrahedral, $C_{2v}$, and square planar geometries. Reproduced with permission from Poliakoff, *Chem. Soc. Rev.*, 7 527 © 1978 Royal Society of Chemistry.

This reaction occurs on irradiation by the output of a Nernst glower functioning as the source of an infrared spectrometer. This finding led to an interesting study of Turner, Poliakoff *et al.* on the infrared laser-induced isomerization of Fe(CO)$_4$ [12].

A statistical mixture of the different isotopomers of Fe($^{12}$C$^{16}$O)$_{4-x}$($^{13}$C$^{18}$O)$_x$ was formed in an argon matrix at 20 K, by ultraviolet photolysis of matrix-isolated Fe(CO)$_5$ which had been 40% enriched with $^{13}$C$^{18}$O. These isotopomers were distinguished by their different infrared absorption bands in the C–O stretching region. In all there are seven possible isotopomers of Fe($^{12}$C$^{16}$O)$_{4-x}$($^{13}$C$^{18}$O)$_x$: two each of Fe($^{12}$C$^{16}$O)$_3$($^{13}$C$^{18}$O) and Fe($^{12}$C$^{16}$O)($^{13}$C$^{18}$O)$_3$ and three of Fe($^{12}$C$^{16}$O)$_2$($^{13}$C$^{18}$O)$_2$. It was found that irradiation of the matrix with the output of a tunable infrared laser at a frequency corresponding to an absorption of one particular isotopomer promoted a selective intramolecular rearrangement as shown below:

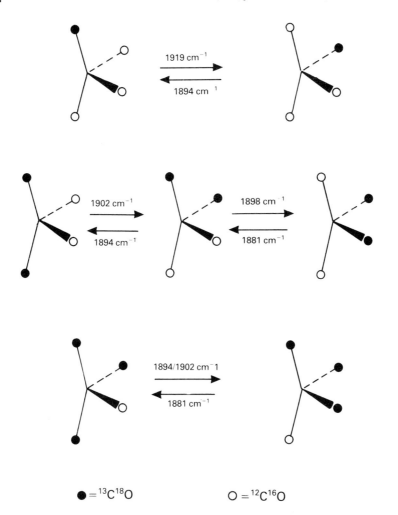

$\bullet = {}^{13}C^{18}O$          $O = {}^{12}C^{16}O$

These interconversions can clearly be described as pseudorotations, but the results show that the permutational mode of this pseudorotation is not that expected for the normal Berry pseudorotation seen, for example, in the thermal rearrangement of $SF_4$. The isomerization of $Fe(CO)_4$ thus provides the first reported example of a non-Berry pseudorotation.

## 3.7  MATRIX-ISOLATION STUDIES ON Mo(CO)₃ AND Mo(CO)₄

As described earlier, ultraviolet photolysis of $Mo(CO)_6$ in a methane matrix produces the 16-electron molecule $Mo(CO)_5$ and free carbon monoxide. Prolonged photolysis, however, causes the sequential loss of further co-ordinated carbonyl groups to form the 14-electron fragment $Mo(CO)_4$ and the 12-electron fragment $Mo(CO)_3$ [13]. Detailed $^{13}C$ isotopic substitution studies have shown that these molecules adopt the following structures (**8, 9**).

(8)                                              (9)

It is interesting to compare the structures of these $d^6$ molybdenum complexes with those of the corresponding $d^8$ iron species. $Fe(CO)_4$ has the $C_{2v}$ structure described in section 4.5 with C–Fe–C angles of 150° and 120°; $Fe(CO)_3$ has a $C_{3v}$ structure with a C–Fe–C angle of 107°. Thus, it appears that the structures of such metal carbonyl fragments may not be determined primarily by Jahn–Teller distortions of more symmetric structures, e.g. $T_d$ for $Mo(CO)_4$, but rather by the maximum overlap possible between filled ligand orbitals and unfilled metal $d$ orbitals. In fact it can be seen that the structures found for these lower carbonyls are only slightly distorted from the structures obtained by selective removal of ligands from an octrahedron.

## 3.8  FORMATION OF THE MOLECULE $Mo(CO)_5N_2$

It has been shown that $Mo(CO)_5$ interacts weakly with noble gases via its empty co-ordination site (see section 3.4) so it is not surprising to find that dinitrogen, $N_2$, co-ordinates to the metal atom of the $Mo(CO)_5$ fragment, forming a complex $Mo(CO)_5N_2$. This may be synthesized by photolysis of $Mo(CO)_6$ in a low-tempera-ture matrix, containing $N_2$, and it has been characterized by its infrared and Raman spectra, polarization measurements having been made on the Raman scattering [14]. The spectral features in the region 1900–2300 cm$^{-1}$ assigned to the molecule $Mo(CO)_5N_2$ are given in Table 3.5, while Fig. 3.6 illustrates the infrared and Raman spectra observed for this molecule.

**Table 3.5** — Vibrational wavenumbers and assignments for the molecule $Mo(CO)_5N_2$

| | $v/cm^{-1}$ | | | | |
|---|---|---|---|---|---|
| Ir absorption | 2253.3 | 2092.8 | — | 1978.2 | 1964.4 |
| Raman scattering | 2253.4 | 2092.3 | 2012.2 | — | 1962.9 |
| State of polarization of Raman lines | polarized | polarized | — | — | polarized |
| Assignment: mode | $v$(N–N) | $v$(C–O) | $v$(C–O) | $v$(C–O) | $v$(C–O) |
| Symmetry class | $a_1$ | $a_1$ | $b_1$ | e | $a_1$ |

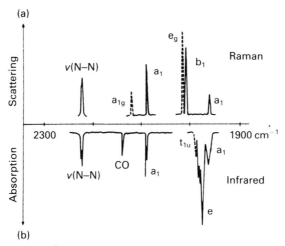

Fig. 3.6 — Infrared and Raman spectra of matrix-isolated Mo(CO)₅N₂. The peaks drawn in dashed lines are due to unreacted Mo(CO)₆. Reproduced with permission from Burdett *et al.*, *Inorg. Chem.*, **17**, 532 © 1978 American Chemical Society.

Polarization studies on Raman lines have allowed assignments to be made to vibrations of the $a_1$ symmetry class. Studies involving isotopic substitution with $^{15}N_2$, $^{14}N^{15}N$ or $^{13}CO$ have confirmed that $Mo(CO)_5N_2$ has $C_{4v}$ symmetry and thus a linear Mo–N≡N unit is implied. The most likely structure for $Mo(CO)_5N_2$ is shown below (**10**).

$$\text{(10)}$$

Unlike the pentacarbonyl species Ar . . . $Mo(CO)_5$, $Mo(CO)_5N_2$ is stable towards irradiation in the visible region of the spectrum. However $Mo(CO)_5N_2$ shows an absorption in the near-ultraviolet region centred at 352 nm in an argon matrix, and irradiation into this band causes both loss of the $N_2$ ligand to form Ar . . . $Mo(CO)_5$ and also uptake of a free CO molecule to regenerate $Mo(CO)_6$. The scheme shown below has been devised for the photochemical interconversion of the three species $Mo(CO)_6$, $Mo(CO)_5N_2$ and Ar . . . $Mo(CO)_5$ in a nitrogen-doped argon matrix.

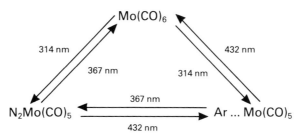

### 3.9   THE PHOTOCHEMISTRY OF $M(CO)_6$ (M = Cr, Mo OR W) MOLECULES IN MATRICES CONTAINING OXYGEN

Photolysis of $M(CO)_6$ molecules in matrices containing oxygen leads to reactions which are rather different from those observed in nitrogen-doped matrices. Instead of the formation of a complex, $O_2M(CO)_5$, which would be analogous to the complex $N_2M(CO)_5$, a number of oxocarbonyl intermediates are produced with fewer than five co-ordinated CO groups, and which also contain oxygen ligands co-ordinated to the metal centre. Ultimately, on prolonged photolysis, all of the carbonyl groups are lost from the metal and the final products of the reaction are binary metal oxides. The intermediate oxocarbonyl complexes, and the metal oxide products, have been the subjects of several interesting experiments.

(11)

It was shown that photolysis of matrix-isolated $Cr(CO)_6$ in the presence of $O_2$ produces, *inter alia*, 'chromyl carbonyl' $Cr(O)_2(CO)_2$ (11) [15]. The infrared spectra of this complex show clearly that the O—O bond has been cleaved by photolysis and it is possible to make estimates of the OC—Cr—CO and O—Cr—O bond angles. The OC—Cr—CO bond angle can be obtained from the intensity ratio of the bands corresponding to the symmetric and antisymmetric carbonyl stretching fundamentals by the relationship:

$$\frac{I_{sym.}}{I_{asym.}} = \tan^2(\theta/2),$$

where $\theta$ is the OC—Cr—CO bond angle. This assumes that the bond dipole model for infrared intensities is valid, that the carbonyl stretching vibrations are effectively uncoupled from all other vibrations in the molecule and that the vibrations are harmonic. Obviously, anharmonicity will cause this estimate of the bond angle to be slightly inaccurate. In principle the O—Cr—O bond angle may also be calculated from intensity measurements though in this case, where Cr—O stretching vibrations are concerned, motion of the chromium atom must be taken into account in the calculations. Moreover, in the case of 'chromyl carbonyl' the $\nu_{sym}(CrO_2)$ vibration was too weak to be observed. However, the O—Cr—O bond angle may also be obtained from the shift in frequency of the asymmetric $CrO_2$ stretching vibration on substitution of the $^{16}O$ atoms, in the molecule, by $^{18}O$. If we define a ratio, $R$, by

$$R = [\nu(^{18}O)/\nu(^{16}O)]^2$$

then an upper limit for the O—Cr—O bond angle, $\theta_u$, may be obtained from the relationship

$$\sin\frac{\theta_u}{2} = \left[\frac{m(Cr)[m(^{16}O) - (m(^{18}O))R]}{2(m(^{16}O))(m(^{18}O))(R-1)}\right]^{1/2}$$

This estimate of the bond angle is an upper limit because of anharmonicity of the vibrations involved.

The molecule $Cr(O)_2(CO)_2$ is of interest because it was the first example of a chromium(IV) carbonyl compound, and yet it appears to be one of the most stable intermediates formed during photo-oxidation of $Cr(CO)_6$. Accordingly, similar experiments have been carried out on the carbonyls $Mo(CO)_6$ and $W(CO)_6$ to look for analogues of 'chromyl carbonyl' [16]. In these experiments, however, the principal oxocarbonyl intermediates observed are the previously unknown complexes, *trans*-dioxotetracarbonylmolybdenum and *trans*-dioxotetracarbonyl-tungsten (**12**).

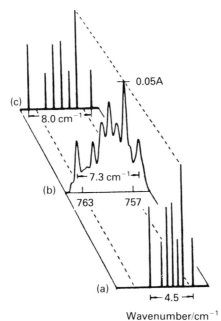

(**12**)

These complexes were characterized by infrared and Raman spectroscopy including isotopic substitution with $^{13}CO$ and $^{18}O$, and observation of splitting of the $\nu_{as}(MoO_2)$ infrared absorption arising from the natural abundance of different isotopes of molybdenum. The spectra showing molybdenum isotope splitting are shown in Fig. 3.7, and they are consistent with presence of a linear O=Mo=O group.

Fig. 3.7 — Molybdenum isotope structure in the infrared spectrum of matrix-isolated *trans*-O₂Mo(CO)₄: (a) spectrum calculated for a monoxo species; (b) observed spectrum; (c) spectrum calculated for a linear dioxo species. Reproduced with permission from Crayston *et al.*, *Inorg. Chem.*, **23**, 3051 © 1984 American Chemical Society.

In order to establish the route by which these oxocarbonyl intermediates are formed from the parent hexacarbonyl molecules, and hence to gain an understanding of the overall reaction mechanism, the following strategy was employed [17]: (i) narrow-band ultraviolet-visible photolysis to build up specific intermediates; (ii) careful monitoring of infrared bands to determine which bands belong to which molecule; (iii) enrichment in $^{13}$CO to determine the number and geometry of co-ordinated CO groups in each intermediate; and (IV) substitution with $^{18}$O to determine the mode of co-ordination of oxygen in each intermediate. Thus, the mechanism shown in Fig. 3.8 has been proposed to account for the formation of $Cr(O)_2(CO)_2$ and *trans*-$M(O)_2(CO)_4$ (M = Mo or W) from $M(CO)_6$ (M = Cr, Mo or W) molecules in $O_2$-doped matrices. A common intermediate in these reactions is the molecule $(O_2) M(CO)_4$ **(13)** which contains an $\eta^2$–$O_2$ (peroxo) ligand with an intact O–O bond and which yields, on near-ultraviolet photolysis, the molecule

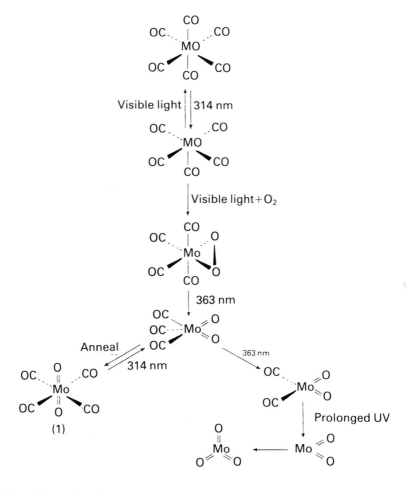

Fig. 3.8 — Scheme showing the photo-oxidation of $Mo(CO)_6$ in an $O_2$-doped Ar matrix. Reproduced with permission from Almond, *Chem. Br.*, **23**, 533 © 1987 Royal Society of Chemistry.

**Table 3.6** — Wavenumbers of infrared absorptions of intermediates in the photo-
oxidation of M(CO)₆ (M = Cr, Mo or W) molecules in argon matrices

| Intermediate | Cr | Mo | W |
|---|---|---|---|
| $(O_2)M(CO)_4$ (**12**) | 2128(vw) | 2133(w) | — |
| | 2062.1(m) | 2064.2(ms) | 2053(m) |
| | 2027.5(ms) | 2016.3(ms) | 2007(m) |
| | 2007.4(m) | 1996(m) | — |
| *trans*-$(O)_2M(CO)_4$ (**11**) | Not observed | 2110.5(s) | 2096(s) |
| $(O)_2M(CO)_3$ (**13**) | 2108(w) | 2106(w) | 2091(m) |
| | 2050(m) | 2025(m) | 1998(m) |
| $(O)_2M(CO)_2$ (**10**) | 2125(s) | 2123(m) | 2112(m) |
| | 2065(s) | 2041(m) | 2018(m) |

$(O)_2M(CO)_3$ (**14**) where the O–O bond has been cleaved. The molecule
$(O)_2M(CO)_3$ may, in turn, act as a precursor to $(O)_2M(CO)_2$ or *trans*-$(O)_2M(CO)_4$
by loss or uptake of a CO molecule. The wavenumbers of the infrared absorptions of
these intermediates are given in Table 3.6. It should be noted that the molecule *trans*-
$(O)_2Cr(CO)_4$ is not observed in these experiments.

(13)                                                    (14)

These reactions provide a route to the production of matrix-isolated binary metal
oxide molecules, on prolonged photolysis, by loss of all co-ordinated carbonyl
groups from the metal centre. Infrared spectroscopy has been used to study these
oxide products, and they have been characterized by isotopic substitution with $^{18}O$.
Chromium hexacarbonyl yields the molecule chromium dioxide, $CrO_2$ [18], while
molybdenum hexacarbonyl forms $MoO_2$ and, on further photolysis, $MoO_3$ [19]. The
formation of binary tungsten oxides from $W(CO)_6$ is, however, somewhat more
complex [19]. The final product is $WO_3$ but this is not produced from $WO_2$. Rather,
the precursor to $WO_3$ appears to be a $WO_2$ molecule, co-ordinated to $O_2$ (**15**), which
forms $WO_3$ by loss of an oxygen atom:

$$(O_2)WO_2 \xrightarrow{h\nu} WO_3 + O$$

(15)

Evidence for the production of oxygen atoms comes from experiments involving oxidation of matrix materials at this point in the reaction. For example, $N_2$ is oxidized to $N_2O$ and methane, $CH_4$, to a mixture of methanol and formaldehyde:

$$N_2 + O \longrightarrow N_2O$$
$$CH_4 + O \longrightarrow CH_3OH$$
$$CH_4 + O \longrightarrow H_2C = O + H_2$$

All three of these reactions are typical of oxygen atoms in the excited $^1D$ state and are known from gas phase studies.

There is interest in these oxidized transition metal molecules because they may be similar in structure to key intermediates in oxidation reactions catalysed by transition metals. Currently, there is great interest in finding transition metal catalysts for the selective oxidation of organic substrates. For example, alkene epoxidation is promoted by $Mo/SiO_2$ heterogeneous catalysts, presumably via oxidized molybdenum species, which may be similar to the intermediates observed in the photo-oxidation of $Mo(CO)_6$ by $O_2$. From an understanding of the structure and bonding in these unstable oxo–metal molecules it will be possible to learn more about intermediates in catalytic oxidation reactions and this may allow the possibility of developing more efficient catalytic systems.

## 3.10 REACTONS OF BINUCLEAR AND TRINUCLEAR METAL CARBONYLS

The photochemistry of binuclear or trinuclear metal carbonyls is likely to be rather more complex than that of mononuclear metal carbonyls. For example, two possible primary photochemical reactions can be envisaged for the molecule $Mn_2(CO)_{10}$. These are cleavage of the Mn–Mn bond to form two $Mn(CO)_5$ radicals, or loss of CO to yield $Mn_2(CO)_9$.

$$(CO)_5Mn-Mn(CO)_5 \longrightarrow 2Mn(CO)_5$$
$$(CO)_5Mn-Mn(CO)_5 \longrightarrow Mn_2(CO)_9 + CO$$

Wrighton and Ginley [20] have shown that photolysis of $Mn_2(CO)_{10}$, $Re_2(CO)_{10}$ or $MnRe(CO)_{10}$ in carbon tetrachloride solution yields as the principal product the corresponding metal pentacarbonyl chloride; for example, $Re_2(CO)_{10}$ undergoes photolysis as follows:

$$Re_2(CO)_{10} \xrightarrow[CCl_{4 \text{ soln.}}]{\lambda = 313 \text{ nm}} 2Re(CO)_5Cl$$

In the presence of $I_2$ the expected $M(CO)_5I$ products are formed in essentially quantitative yields. This photochemistry is consistent with homolytic metal–metal bond cleavage arising from a $\sigma \rightarrow \sigma^*$ one-electron transition associated with the metal–metal bond.

However, more recent experiments have been carried out in which $Mn_2(CO)_{10}$ is subjected to laser flash-photolysis in cyclohexane solution [21]. These studies suggest that both Mn–Mn and Mn–CO bond breaking occurs. In the presence of the phosphine, $P(n\text{-Bu})_3$, the product $Mn_2(CO)_9 P(n\text{-Bu})_3$ is observed and from the kinetics of this reaction it appears that this product is formed via the intermediate $Mn_2(CO)_9$ in a two-step process:

$$Mn_2(CO)_{10} \longrightarrow Mn_2(CO)_9 + CO$$
$$Mn_2(CO)_9 + P(n\text{-}Bu)_3 \longrightarrow Mn_2(CO)_9P(n\text{-}Bu)_3$$

rather than via $Mn(CO)_5^\bullet$, involving a more complex reaction sequence, for example:

$$Mn_2(CO)_{10} \longrightarrow 2Mn_2(CO)_5^\bullet$$
$$Mn(CO)_5^\bullet + P(n\text{-}Bu)_3 \longrightarrow Mn(CO)_4P(n\text{-}Bu)_3^\bullet + CO$$
$$Mn(CO)_4[P(n\text{-}Bu)_3]^\bullet + Mn(CO)_5^\bullet \longrightarrow Mn_2(CO)_9P(n\text{-}Bu)_3 .$$

Following these studies, experiments were carried out in which $Mn_2(CO)_{10}$, trapped in methylcyclohexane or 3-methylpentane glasses at 77 K [22], or in argon matrices at 12 K [23], was subjected to photolysis. In these rigid media, CO loss is seen as the sole discernible primary photolysis step, presumably because $Mn(CO)_5^\bullet$ radicals readily recombine in the matrix cage. The matrix studies of Dunkin et al. [23] involved the use of polarized radiation for photolysis and polarized spectroscopy. These experiments demonstrate that an infrared band at $1764 \text{ cm}^{-1}$, assigned to a bridging carbonyl stretching mode, shows no dichroism, implying that the group responsible for this absorption is either fluxional, or is oriented at an angle of approximately 45° from the Mn–Mn bond. The latter explanation appears to be more likely, and is consistent with a semi-bridged carbonyl structure of $Mn_2(CO)_9$ (**16**).

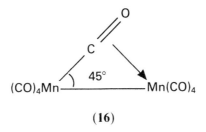

(**16**)

The photochemical reactions of $Re_2(CO)_{10}$ in argon and nitrogen matrices have recently been studied [24]. On photolysis in argon matrices, two forms of $Re_2(CO)_9$ are seen. These are equatorial (**17**) and axial (**18**) isomers, and they can be interconverted by UV-visible photolysis.

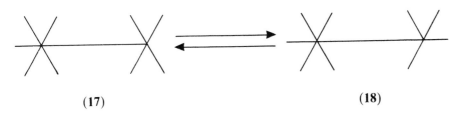

(**17**)                                                  (**18**)

In nitrogen matrices, at a temperature of 10 K, $Re_2(CO)_9$ is likewise formed by UV photolysis. On annealing the matrix to 15–20 K, the isomer $eq - Re_2(CO)_9(N_2)$ (**19**) is generated by a bimolecular reaction. Photolysis of the matrix with visible light ($\lambda = 546$ nm) converts **19** to the axial isomer, $ax\text{-}Re_2(CO)_9(N_2)$, (**20**), while **20** can be reconverted to **19** by irradiation at 313 nm. These complex photochemical reactions have been interpreted with the help of results from solution studies, in particular studies in low-temperature liquid xenon solution.

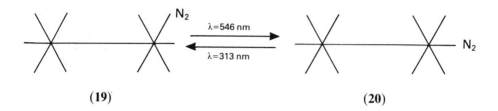

$\lambda=546$ nm

$\lambda=313$ nm

N$_2$

(**19**)          (**20**)

Finally, the photochemistry of $Mn_2(CO)_{10}$ in $O_2$-doped argon matrices had been investigated [25]. Under these conditions it is found that the ultimate product of photolysis is the O-bridged molecule $Mn_2O_7$ (**21**). This complicated reaction involves breaking of the Mn–Mn bond, loss of all CO groups from $Mn_2(CO)_{10}$ and formation of an Mn–O–Mn bridge. However, there is, as yet, little evidence concerning the nature of any intermediate species. By contrast, photo-oxidation of $Re_2(CO)_{10}$ yields, as the ultimate reaction product, a binary oxide with no bridging oxygen [25].

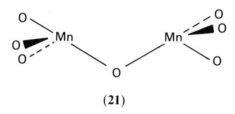

(**21**)

Reactions of trinuclear carbonyls are likely to be even more complicated than those of binuclear carbonyls. However, some work has been carried out on the molecules $Fe_3(CO)_{12}$ and $Ru_3(CO)_{12}$ [26]. In methylcyclohexane glass at 90 K, these molecules show CO loss as the primary photolytic step yielding $M_3(CO)_{11}$ molecules (M = Fe or Ru). The reactions are complicated, however, by the presence of different isomers of the products; for example, an axially vacant form of $Ru_3(CO)_{11}$, having no bridging carbonyl groups, rearranges, at 90 K, to a form having at least one bridging CO group. In hydrocarbon solution, in the presence of ethylene, photofragmentation of the trinuclear cluster occurs.

$$Ru_3(CO)_{12} \xrightarrow[C_2H_4]{h\nu\ 195\ K} Ru(CO)_4(C_2H_4) + Ru_2(CO)_8(C_2H_4)$$

These findings have important implications for catalytic systems. In particular, alkene isomerization catalysts may be formed by low-energy visible light excitation of $Fe_3(CO)_{12}$ or $Ru_3(CO)_{12}$, and key intermediates in these catalytic systems may well be akin to species observed in solution, or in low-temperature glasses.

## 3.11  STUDIES OF RELATED SYSTEMS

At this point it is interesting to consider, briefly, the photochemistry of some complexes which are related to metal carbonyls. Two classes of complex will be considered. These are the metal nitrosyls and ethylene complexes of metals.

The photochemistry of the mixed carbonyl nitrosyl complex [V(Cp)(CO)(NO)$_2$] (where Cp = $\eta^5$-cyclopentadienyl) has been studied in low-temperature matrices [27]. It is found that ultraviolet irradiation of this complex causes CO loss as the primary photochemical step, yielding the complex [V(Cp)(NO)$_2$], This reaction illustrates the fact that, in general, carbonyl groups are more photolabile than nitrosyl groups. Thus for example, the complexes [Mn(NO)(CO)$_4$], [Mn(NO)$_3$(CO)] and [Fe(NO)$_2$(CO)$_2$] all ʻhow CO loss as the primary photoreaction, in argon or methane matrices [28]. However, matrix-isolated [V(Cp)(CO)(NO)$_2$] shows a rather different reaction, if it is subjected to *visible* iradiation. In this case, CO is not lost from the metal centre. Rather, one of the NO groups is converted from a three-electron ligand to a one-electron ligand, and is identified by a very low v(N–O) stretching frequency. This implies that the co-ordination of NO has changed from linear (**22**) to non-linear (**23**) on visible

(**22**)                                              (**23**)

three-electron donor                         one electron donor

photolysis. The infrared absorptions of matrix-isolated [V(Cp)(CO)(NO)$_2$] and [V(Cp)(CO)(NO)(NO$^*$)] (where NO$^*$ represents NO acting as a one-electron donor) are listed in Table 3.7.

**Table 3.7** — Infrared absorptions of the complexes
[V(Cp)(CO)(NO)$_2$] and [V(Cp)(CO)(NO)(NO$^*$)]
isolated in methane matrices at 12 K

| Complex | v/cm$^{-1}$ | Assignment |
|---|---|---|
| [V(Cp)(CO)(NO)$_2$] | 2054.0 | v(CO) |
| | 1731.0 | v(NO) |
| | 1649.2 | v(NO) |
| [V(Cp)(CO)(NO)(NO$^*$)] | 2080.0 | v(CO) |
| | 1652.9 | v(NO) |
| | 1393.9 | v(NO$^*$) |

An interesting study of the photolysis of the complex [Rh(NO)(CO)(PPh$_3$)$_2$] in dichloromethane solution has shown that NO is expelled rather than CO, and that the product is *trans*-[Rh(CO)Cl(PPh$_3$)$_2$] [29]. It is proposed that photodissociation of

NO proceeds from a charge-transfer state with a bent M–N–O moiety (23), in which the NO acts as a one-electron donor. Thus the M–N bond is considerably weakened, facilitating dissociation of NO.

Ethylene complexes of transition metals are generally highly photolabile. Photolysis of the complex $[(Cp)_2W(C_2H_4)]$ (24) in an argon matrix at 12 K [30], results in loss of the $C_2H_4$ ligand and formation of the 'sandwich' compound tungstenocene, $[(Cp)_2W]$, (25). More recently, it has been shown [31] that photolysis of the complex $[(Cp)Rh(C_2H_4)_2]$ (26) in an argon matrix leads to loss of $C_2H_4$ and formation of $[(Cp)Rh(C_2H_4)]$ (27). In a CO matrix, 27 will in turn take up CO to form $[(Cp)Rh(C_2H_4)(CO)]$ (28) but this complex is itself photolabile with respect to $C_2H_4$ loss, and further photolysis, in a CO matrix, results in the formation of $[(Cp)Rh(CO)_2]$ (29). These reactions are summarized below. Fig. 3.9 illustrates the infrared spectra seen when $[(Cp)Rh(C_2H_4)_2]$ (25) is photolysed in a methane matrix. These spectra show clearly the formation of $[(Cp)Rh(C_2H_4)]$ (26), alongside free unco-ordinated ethylene.

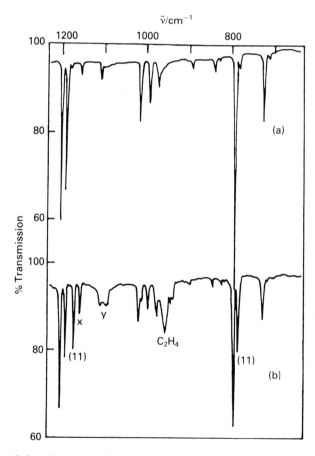

Fig. 3.9 — Infrared spectra of CpRh(C₂H₄)₂ in a methane matrix: (a) before photolysis; (b) after broad-band photolysis. Bands marked (11) are due to CpRh(C₂H₄). Reproduced with permission from Haddleton and Perutz, *J. Chem. Soc., Chem. Commun.*, 1372 © 1985 Royal Society of Chemistry.

These reactions are of particular interest because some of these intermediates are potent activators of the C–H bonds of alkanes. More discussion of these reactions will be given in section 4.5.

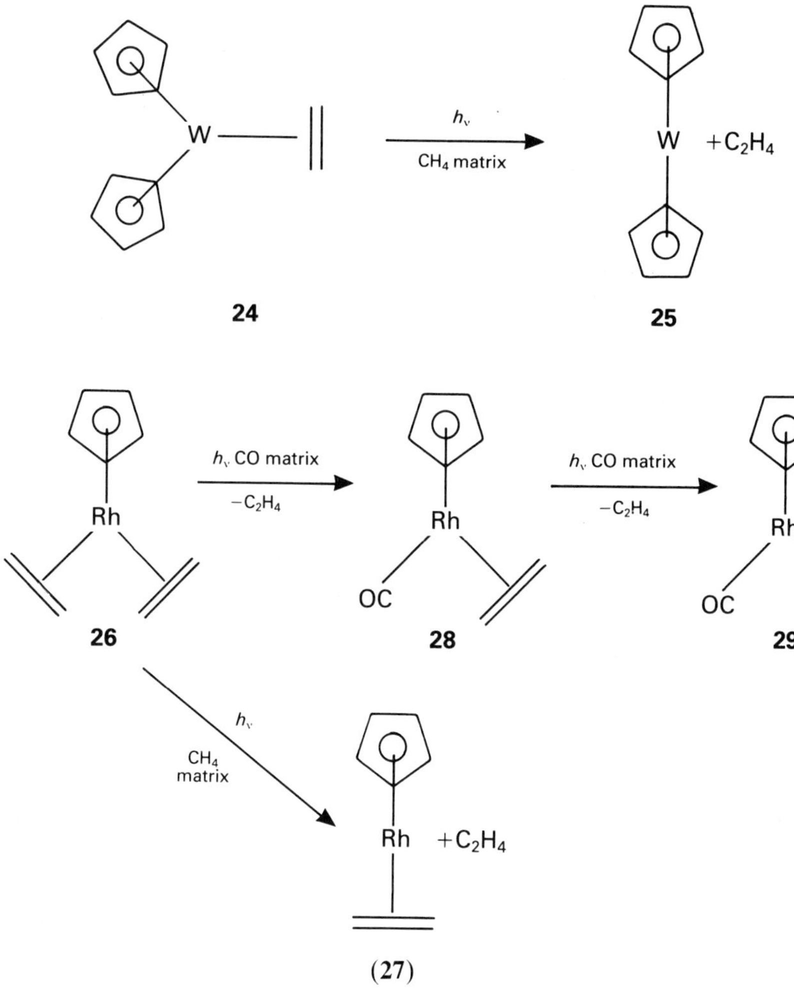

**3.12  STUDIES IN LIQUID NOBLE GAS SOLUTIONS**

It is always desirable to complement studies carried out in low-temperature matrices by experiments in rather more conventional media. A connection between, on the one hand, results obtained from solid, low-temperature matrices, and on the other, those derived from non-polar hydrocarbon solvents, has been made by the work carried out in low-temperature *liquid* noble gases. Such systems have two main advantages over conventional matrix-isolation and solution experiments. In the first case, in these fluid media, kinetic measurements can be made, and thus, at least in principle, estimates can be made of bond energies. Secondly very long path lengths can be employed for infrared spectroscopic study of the solutions, since the noble gas solvents themselves show no infrared absorptions. Thus very weak infrared bands of the species in solution may be detected, which might otherwise have escaped observation. Two recent studies by Poliakoff, Turner and their co-workers at the University of Nottingham clearly illustrate these points.

The molecule $Mo(CO)_5(N_2)$ was discussed in section 3.7, and while the structure of this complex was obtained from a spectroscopic study of low-temperature matrices, there was no indication as to the thermal stability of the molecule, nor of the strength of the $M-N_2$ bond. However, recent work has shown that photolysis of $Cr(CO)_6$ in liquid xenon solutions containing dinitrogen yields $Cr(CO)_5(N_2)$, and that this product is able to survive at temperatures as high as $-35°C$ [32]. In this way other members of the series $Cr(CO)_{6-x}(N_2)_x$ $(x=2-5)$ have been identified, and it is found that the thermal stability decreases as $x$ increases. In a similar manner the transient $Ni(CO)_3(N_2)$ has been generated by photolysis of $Ni(CO)_4$ in liquid krypton doped with dinitrogen at 114 K [33]. This species is thermally unstable, under these conditions, and decays slowly to regenerate $Ni(CO)_4$. It is possible, from a study of the kinetics of the decay as a function of temperature, to establish kinetic and energetic parameters for the reaction

$$Ni(CO)_3(N_2)+CO \longrightarrow Ni(CO)_4+N_2$$

Hence an estimate of about 40 kJ mol$^{-1}$ for the activation enthalpy of the dissociative pathway gives a measure of the $Ni-N_2$ bond energy in $Ni(CO)_3(N_2)$.

There has recently been much interest in the reactions between transition metal carbonyls and dihydrogen. Sweany discovered that a reversible interconversion between $Fe(CO)_4$ and $Fe(CO)_4(H_2)$ could be initiated in an $H_2$-doped argon matrix by selective photolysis.

$$Fe(CO)_4 \underset{\text{UV visible}}{\overset{\text{infrared}}{\rightleftharpoons}} Fe(CO)_4(H_2)$$

In this case it appears that hydrogen is co-ordinated in the classical dihydride manner (**30**). However, when Sweany carried out similar experiments, in which he photolysed $Cr(CO)_6$ in $H_2$-doped argon matrices, he observed *inter alia* the product $Cr(CO)_5 H_2$ and, in this case, there was a strong body of circumstantial evidence in favour of the dihydrogen structure, **31**, in which the H–H bond remains intact [34]. Unfortunately the weak H–H stretching mode could not be observed, but this

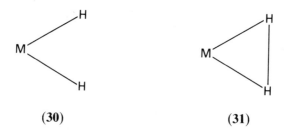

(**30**)                                        (**31**)

problem was overcome by Poliakoff, Turner and their co-workers who repeated the experiment in liquid xenon solution at 200 K [35]. Taking advantage of the long path length (27 mm) of their cell, these workers were able to observe the absorption corresponding to the H–H stretching mode and its D–D counterpart when $D_2$ was substituted for $H_2$ as the solution dopant. These infrared spectra are illustrated in Fig. 3.10.

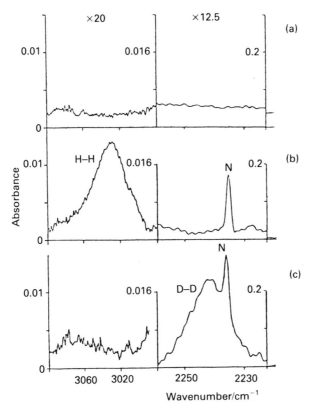

Fig. 3.10 — Infrared spectra of $Cr(CO)_6$ in liquid xenon solution: (a) before photolysis; (b) after photolysis in $H_2$-doped Xe; (c) after photolysis in $D_2$-doped Xe. N is an impurity band. Reproduced with permission from Upmacis *et al.*, *J. Chem. Soc., Chem. Commun.*, 27 © 1985 Royal Society of Chemistry.

At first sight a complex with the structure **31** may appear to be extraordinary, but the most likely explanation for the existence of such a complex is that the bonding of $H_2$ to the metal centre consists of charge donation from a filled metal $d$ orbital into the antibonding, $\sigma^*$, orbital of the $H_2$ ligand [32]. This weakens, but does not rupture the H–H bond.

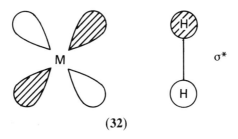

**(32)**

These examples show in particular how matrix-isolation results can be supplemented by studies in liquid noble gas solutions and also how the solution work can provide answers to problems which are difficult, or impossible, to solve by matrix-isolation experiments alone.

### 3.13   TIME-RESOLVED INFRARED SPECTROSCOPIC STUDIES

This section concentrates on another type of experiment which has become increasingly important in recent years and the results from which complement earlier matrix-isolation studies. This is the time-resolved infrared spectroscopic study of reactive species in room-temperature solution — a variation on classical flash photolysis experiments. The reactive species are generated by a photolytic flash from a pulsed ultraviolet source. Kinetic measurements are then made on the sample at one infrared frequency by means of a tunable infrared laser. The monitoring infrared frequency is then changed and the flash activated again. Thus, by using a number of flashes, data are accumulated at frequencies across the spectral region of interest, for example, the C–O stretching region. Like classical flash photolysis studies these experiments provide kinetic information, but the use of infrared radiation to monitor the sample gives added structural information and helps to identify more unambiguously the reaction intermediates. One recent example will illustrate the great value of this technique.

In 1983, Rest and his co-workers at Southampton described the results of photolysing the dimer $[CpFe(CO)_2]_2$ ($Cp = \eta^5$-$C_5H_5$) (33) in a methane matrix at 12 K [36]. The product of this reaction is the CO-bridged dimer (34), the structure of which has been confirmed by $^{13}CO$ substitution. When 33 is subjected to flash photolysis in cyclohexane solution at room temperature, time-resolved infrared studies show that both 34 and the radical $[CpFe(CO)_2]^\bullet$ are formed within 5 $\mu s$ [37].

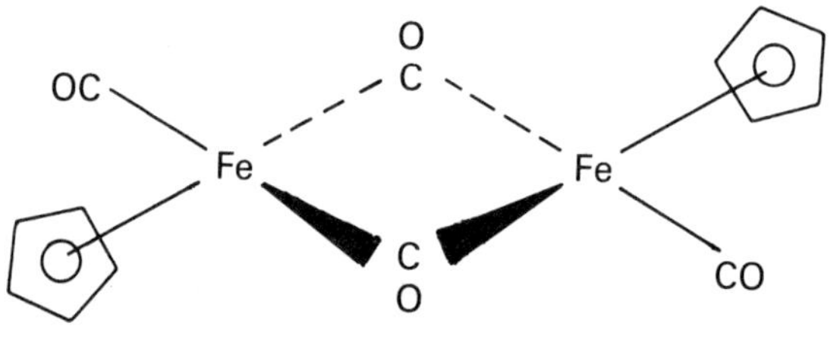

(33)

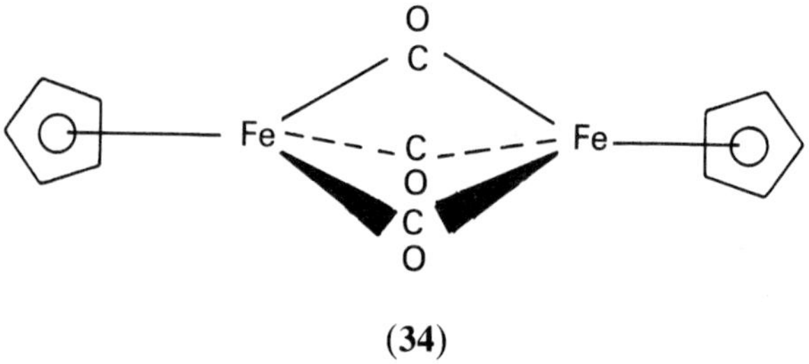

(34)

The radical has a half-life of <25 $\mu$s, whereas **34** has a half-life of 1.5 ms under the experimental conditions. If the cyclohexane solution is doped with 2-electron donor ligands other than CO, e.g. $L = CH_3CN$ or $PPh_3$, it is possible to monitor the formation of the substitution product $Cp_2Fe_2(CO)_3L$:

$$[CpFe(CO)_2]_2\,L \xrightarrow[\substack{cyclohexane \\ solution}]{UV} Cp_2Fe_2(CO)_3L + CO$$

Kinetic studies, as illustrated in Fig. 3.11, leave little doubt that the principal intermediate in photo-substitution is not the radical but the binuclear species **34**. Hence the rate constant for the reaction, when $L = CH_3CN$, is found to be $7.6 \times 10^5$ $dm^3\,mol^{-1}\,s^{-1}$ at 24°C and the activation energy to be $24.4 \pm 1$ kJ mol$^{-1}$. However,

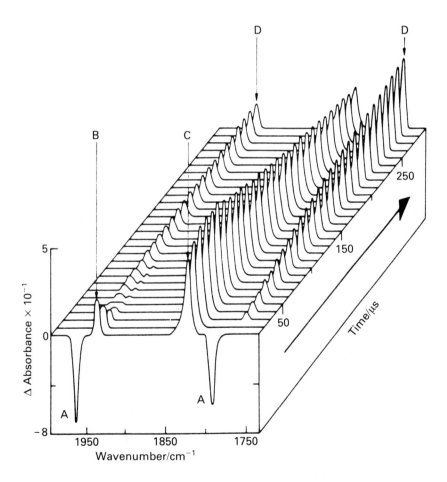

Fig. 3.11 — Time-resolved infrared spectrum obtained after flash photolysis of [CpFe(CO)$_2$] and MeCN in cyclohexane solution. Two intermediates, B = CpFe(CO)$_2'$, and C = CpFe($\mu$-CO)$_3$FeCp are seen. D is the reaction product Cp$_2$Fe$_2$(CO)$_3$(MeCN). Reproduced with permission from Dixon *et al.*, *J. Chem. Soc., Chem. Commun.*, 994 © 1986 Royal Society of Chemistry.

more recent studies [38] have shown that the radical $CpFe(CO)_2^{\cdot}$ reacts with the phosphine $P(OMe)_3$, in $n$-heptane solution via

$$CpFe(CO)_2 + P(OMe)_3 \longrightarrow CpFe(CO)P(OMe)_3^{\cdot} + CO$$

an associative pathway with a bimolecular rate constant of $8.9 \pm 2.0 \times 10^8 \, dm^3 \, mol^{-1} \, s^{-1}$ at 25°C. This is only about one order of magnitude slower than the diffusion-controlled limit.

What is clear from these experiments is that solution studies can provide kinetic data, which it is impossible to derive from matrix-isolation studies. Thus they provide much more information about the *reactivity* of short-lived intermediates, and give a valuable insight into reaction mechanisms. Matrix isolation remains, however, the best approach to the structural characterization of transient intermediates, and so the two techniques readily complement each other. The solution study on $[CpFe(CO)_2]_2$, described above, would have been difficult to interpret without Rest's initial characterization of the intermediate $Cp_2Fe_2(\mu\text{-}CO)_3$ (**34**) in a low-temperature matrix.

### 3.14  GAS PHASE STUDIES

This chapter on metal carbonyl photochemistry ends with a brief description of some of the experiments which have been carried out on metal carbonyls in the gas phase. To some extent the results from these experiments mirror the results of matrix-isolation studies on the same systems. Thus, pulsed laser excitation of vapour phase $Cr(CO)_6$ at 249 nm yields $Cr(CO)_5$ and CO [39], the same products as are seen on matrix ultraviolet photolysis. The reaction has been followed in real time by time-resolved infrared laser spectroscopy, a method which allows the concentrations both of reactants and of all major products to be followed. In this way, two important differences between the behaviour of $Cr(CO)_6$ on ultraviolet photolysis in the gas phase and in a low-temperature matrix have emerged. The first is the relatively inefficient energy transfer in low-pressure gas phase systems. This leads to rapid decomposition of $Cr(CO)_5$ to $Cr(CO)_4$ and CO, because of the high internal energy of the photolytically generated $Cr(CO)_5$. Indeed on photolysis at 249 nm, $Cr(CO)_4$ is formed within $10^{-7}$ s of the activating laser pulse. Secondly, bimolecular reactions occur more readily in the gas phase. Thus the processes

$$Cr(CO)_4 + CO \longrightarrow Cr(CO)_5 ,$$
$$Cr(CO)_5 + CO \longrightarrow Cr(CO)_6$$

and

$$Cr(CO)_4 + Cr(CO)_6 \longrightarrow Cr_2(CO)_{10}$$

all take place. The last of these reactions involving two relatively large molecules would be very unlikely to take palce under matrix conditions. In the gas phase, however, it appears that $Cr(CO)_4$ and $Cr(CO)_6$ react without an activation barrier, although the structure of the $Cr_2(CO)_{10}$ product remains unclear. Two possibilities are a CO bridged structure (**35**) or an M–M bonded structure (**36**).

$$
\begin{array}{c}
O \\
\parallel \\
C
\end{array}
$$

(CO)$_4$Cr          Cr(CO)$_4$

$$
\begin{array}{c}
C \\
\parallel \\
O
\end{array}
$$

(CO)$_5$Cr————Cr(CO)$_5$

(**35**)                                              (**36**)

The co-ordinatively unsaturated $Cr(CO)_x$ species ($x=2$–5) have all been detected by time-resolved infrared spectroscopy, following production in the gas phase by excimer laser photolysis of $Cr(CO)_6$ [40]. The results from these experiments indicate that the gas phase structures of the $Cr(CO)_x$ molecules are the same as the structures observed for the matrix-isolated molecules, Thus $Cr(CO)_5$ adopts a $C_{4v}$ structure (**2**), $Cr(CO)_4$ a $C_{2v}$ structure (**8**) and $Cr(CO)_3$ a $C_{3v}$ structure (**9**).

(**2**)                     (**8**)                     (**9**)

The formation of these different chromium carbonyl fragments depends on the wavelength of excitation. Thus $Cr(CO)_5$ is seen as the principal product 0.5 $\mu s$ after irradiation of $Cr(CO)_6$ at 351 nm. However, 0.5 $\mu s$ after irradiation at 248 nm, $Cr(CO)_4$ is the principal product, and a high yield of $Cr(CO)_3$ is seen when the exciting radiation is at 193 nm. These results are illustrated by the spectra in Fig. 3.12, and can be explained in terms of the shorter wavelength radiation being sufficiently energetic to remove more than one CO group.

Finally it is interesting to consider the reactions of some gas phase metal carbonyl anions with dioxygen [41]. Metal carbonyl anions, for example $Cr(CO)_5^-$ and $Mo(CO)_5^-$, can be formed by electron impact on the parent carbonyls $Cr(CO)_6$ and $Mo(CO)_6$, and they can be made to react with $O_2$ in a fast flow of helium gas. Products have been detected by mass spectrometry, and include $[Cr(CO)_3O_2]^-$,

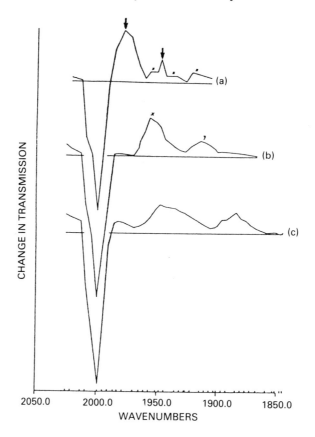

Fig. 3.12 — Photolysis of $Cr(CO)_6$ in the gas phase: (a) at $\lambda = 351$ nm; (b) at $\lambda = 248$ nm; (c) at $\lambda = 193$ nm. Reproduced with permission from Seder *et al.*, *J. Am. Chem. Soc.*, **108**, 4721 © 1986 American Chemical Society.

$[Mo(CO)_3O_2]^-$ and $[Mo(CO)_4O_2]^-$. The structure **37** has been proposed for the dioxometal tricarbonyl intermediates.

$$
\left[
\begin{array}{c}
\end{array}
\right]^{-}
$$

(**37**)

Interestingly the ion $[Cr(CO)_3O_2]^-$ appears to be relatively inert towards further oxidation, whereas $[Mo(CO)_3O_2]^-$ and $[Mo(CO)_4O_2]^-$ readily give way to $MoO_x^-$ ions if the concentration of $O_2$ in the gas flow is increased. Such information may have implications for the matrix-isolation studies on the photo-oxidation of metal carbonyls described in section 3.9.

**3.15  SUMMARY**

In this chapter I have presented a summary of only a small part of the wealth of interesting chemistry displayed by unstable intermediates in reactions of metal carbonyls. To a large extent I have concentrated on the technique of matrix isolation, and in part this results from the ideal nature of metal carbonyls as subjects for matrix-isolation studies. In the main, such substances are volatile enough to be readily incorporated into an inert gas matrix, and the molecules are generally highly photosensitive, readily yielding reactive intermediates. Moreover, the detection and characterization of metal carbonyls by vibrational spectroscopy are often relatively straightforward tasks. The vibrational modes involving C–O stretching are normally strong in infrared absorption and Raman scattering, allowing easy detection, even in a very dilute sample and the position of these bands is very sensitive to the chemical environment of the metal centre. Thus a large number of matrix-isolation studies have been devoted to transition metal carbonyl chemistry.

However, while matrix isolation is an excellent technique for structure determination it does not tell us much about the kinetic reactivity of transient intermediates trapped in the matrix. Thus modern developments in fast time-resolved infrared spectroscopy assume a great importance. It is now possible to measure the infrared spectra of transient intermediates in solution or in the gas phase on a nanosecond timescale. The results of these experiments will complement matrix-isolation studies by providing important kinetic information on species which had hitherto been observed only when trapped at low temperatures.

Finally, studies in low-temperature solutions are becoming more important. In this work, reactive intermediates are 'slowed down' rather than 'stopped' as in a matrix-isolation experiment. Thus, while conventional spectroscopic techniques may be used to monitor the species present, kinetic information can still be obtained, for example by monitoring the rate of decomposition of a particular complex at different solution temperatures.

The use of these three techniques, in conjunction, should ensure a healthy future for this area of chemistry and lead to the discovery of many more interesting reactions of transition metal carbonyls.

**REFERENCES**

[1]  G. R. Dobson, M. A. El Sayed, I. W. Stolz and R. K. Sheline, *Inorg. Chem.* (1962), **1**, 526

[2]  I. W. Stolz, G. R. Dobson and R. K. Sheline, *Inorg. Chem.* (1963), **2**, 1264.

[3]  J. Nasielski, P. Kirsch and L. Wilputte-Steinert, *J. Organomet. Chem.* (1971), **29**, 269.

[4]  J. M. Kelly, H. Hermann and E. Koerner von Gustorf, *J. Chem. Soc., Chem. Commun.* (1973), 105.

[5]  I. W. Stolz, G. R. Dobson and R. K. Sheline, *J. Am. Chem. Soc.* (1963), **85**, 1013.

[6]  J. D. Black and P. S. Braterman, *J. Organomet. Chem.* (1973), **63**, C19.

[7]  M. A. Graham, M. Poliakoff and J. J. Turner, *J. Chem. Soc. A*, (1971), 2939.

[8]  R. N. Perutz and J. J. Turner, *Inorg. Chem.* (1974), **14**, 262.

[9]   J. K. Burdett, J. M. Grzybowski, R. N. Perutz, M. Poliakoff, J. J. Turner and R. F. Turner, *Inorg. Chem.* (1978), **17**, 146.

[10]  M. Poliakoff, *Chem. Soc. Rev.* (1978), **7**, 527.

[11]  J. K. Burdett, *J. Chem. Soc., Faraday Trans. II* (1974), **70**, 1599.

[12]  B. Davies, A. McNeish, M. Poliakoff and J. J. Turner, *J. Am. Chem. Soc.* (1977), **99**, 7573.

[13]  R. N. Perutz and J. J. Turner, *J. Am. Chem. Soc.* (1975), **97**, 4800.

[14]  J. K. Burdett, A. J. Downs, G. P. Gaskill, M. A. Graham, J. J. Turner and R. F. Turner, *Inorg. Chem.* (1978), **17**, 523.

[15]  M. Poliakoff, K. P. Smith, J. J. Turner and A. J. Wilkinson, *J. Chem. Soc., Dalton Trans.* (1982), 651.

[16]  J. A. Crayston, M. J. Almond, A. J. Downs, M. Poliakoff and J. J. Turner, *Inorg. Chem.* (1984), **23**, 3051.

[17]  M. J. Almond, J. A. Crayston, A. J. Downs, M. Poliakoff and J. J. Turner, *Inorg. Chem.* (1986), **25**, 19.

[18]  M. J. Almond and M. Hahne, *J. Chem. Soc., Dalton Trans.* (1988), 2255.

[19]  M. J. Almond and A. J. Downs, *J. Chem. Soc., Dalton Trans.*, (1988), 809.

[20]  M. S. Wrighton and D. S. Ginley, *J. Am. Chem. Soc.* (1975), **97**, 2065.

[21]  H. Yesaka, T. Kobayashi, K. Yasufuku and S. Nagakura, *J. Am. Chem. Soc.* (1983), **105**, 6249.

[22]  A. F. Hepp and M. S. Wrighton, *J. Am. Chem. Soc.* (1983), **105**, 5934.

[23]  I. R. Dunkin, P. Härter and C. J. Shields, *J. Am. Chem. Soc.* (1984), **106**, 7248.

[24]  S. Firth, W. E. Klotzbücher, M. Poliakoff and J. J. Turner, *Inorg. Chem.* (1987), **26**, 3370.

[25]  M. J. Almond, *J. Mol. Struct.* (1988) **172**, 157; M. J. Almond and R. H. Orrin, unpublished results.

[26]  J. G. Bentsen and M. S. Wrighton, *J. Am. Chem. Soc.* (1987), **109**, 4530.

[27]  M. Herberhold, W. Kremnitz, H. Trampisch, R. B. Hitam, A. J. Rest and D. J. Taylor, *J. Chem. Soc., Dalton Trans.* (1982), 1261.

[28]  O. Crichton and A. J. Rest, *J. Chem. Soc., Dalton Trans.* (1977), 202, 208, 536.

[29]  M. Kubota, M. K. Chan, D. C. Boyd and K. R. Mann, *Inorg. Chem.* (1987), **26**, 3261.

[30]  J. Chetwynd-Talbot, P. Grebenik and R. N. Perutz, *Inorg. Chem.* (1982), **21**, 3647.

[31]  D. M. Haddleton and R. N. Perutz, *J. Chem. Soc., Chem. Commun.* (1985), 1372.

[32]  J. J. Turner, M. B. Simpson, M. Poliakoff, W. B. Maier, II and M. A. Graham, *Inorg. Chem.* (1983), **22**, 911.

[33]  J. J. Turner, M. B. Simpson, M. Poliakoff and W. B. Maier, II, *J. Am. Chem. Soc.* (1983), **105**, 100.

[34]  R. L. Sweany, *J. Am. Chem. Soc.* (1985), **107**, 2374.

[35]  R. K. Upmacis, G. E. Gadd, M. Poliakoff, M. B. Simpson, J. J. Turner, R. Whyman and A. F. Simpson, *J. Chem. Soc., Chem. Commun.* (1985), 27.

[36]  R. H. Hooker, K. A. Mahmoud and A. J. Rest. *J. Chem. Soc., Chem. Commun.* (1983), 1022.

[37]  B. D. Moore, M. Poliakoff, M. B. Simpson and J. J. Turner, *J. Phys. Chem.* (1985), **89**, 850.

[38] A. J. Dixon, S. J. Gravelle, L. J. van de Burgt, M. Poliakoff, J. J. Turner and E. Weitz, *J. Chem. Soc., Chem. Commun.* (1987), 1023.

[39] T. R. Fletcher and R. N. Rosenfeld, *J. Am. Chem. Soc.* (1985), **107**, 2203.

[40] T. A. Seder, S. P. Church and E. Weitz, *J. Am. Chem. Soc.* (1986), **108**, 4721.

[41] K. Lane, L. Sallans and R. R. Squires, *J. Am. Chem. Soc.* (1984), **106**, 2719.

# 4

# Reactivity of metal atoms

## 4.1 INTRODUCTION

Metal atoms undergo a variety of interesting chemical reactions and many such reactions have been investigated by the matrix-isolation technique. Thus, co-condensation of metal atoms — produced by high-temperature evaporation of a metal in a Knudsen cell, by laser vaporization of a metal, or by sputtering of metal filaments — with various reactive molecules and an excess of inert gas leads to matrix-isolated products which can be characterized spectroscopically. Often these reactions involve simple addition; for example, many metal atoms will add to carbon monoxide to form metal carbonyls. But various abstraction and insertion reactions are also known. Some metal atoms will abstract an oxygen atom from nitrous oxide, $N_2O$, to form metal oxides, while various insertion reactions of metal atoms into $C-H$, $C-C$ and other bonds have been observed. Several of these reactions have important implications for catalytic processes.

Aggregation reactions of metals to produce dimers and small metal clusters have been studied either in low-temperature matrices, or in supersonic molecular beams. The interest in small metal clusters comes, in part, from the connection between isolated clusters and the small metal particles found on the surface of heterogeneous catalysts. A study of the structure and reactivity of isolated metal clusters may therefore help in understanding some of the reactions which take place on catalytic surfaces. The following sections describe areas where fruitful research has been carried out on the reactivity of metal atoms and of small metal clusters.

## 4.2 FORMATION OF METAL CARBONYLS

An example of the reactivity of metal atoms with carbon monoxide is provided by the reaction of nickel atoms. All four binary carbonyls of nickel, $Ni(CO)_{1-4}$, may be produced by co-condensation of nickel atoms with CO and excess argon at 4.2 K [1]. In a 500:1 Ar:CO matrix, at 4.2 K, the principal product, seen on deposition, is the monocarbonyl Ni(CO). However, the concentration of this species decreases if the matrix is annealed, and at the same time infrared absorptions of the molecules $Ni(CO)_2$, $Ni(CO)_3$ and $Ni(CO)_4$ appear and grow. These nickel carbonyl complexes have been identified by means of isotopic substitution with $C^{18}O$. The changes seen in the $C-O$ stretching region of an argon matrix on progressive warming from 4.2 to 35 K are shown in Fig. 4.1. From these spectra it is clear that the following aggregation processes are taking place:

$$Ni + CO \xrightarrow[4.2\,K]{Ar\,matrix} Ni(CO) \xrightarrow[17\,K]{+CO} Ni(CO)_2$$

$$19\,K \downarrow +CO$$

$$Ni(CO)_4 \xleftarrow[26\,K]{+CO} Ni(CO)_3$$

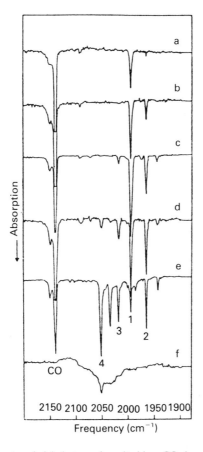

Fig. 4.1 — Infrared spectra of nickel atoms deposited in a CO-doped Ar matrix, followed by annealing: (a) after deposition; (b) 17 K; (c) 18 K; (d) 19 K; (e) 26 K; (f) 35 K. The numbers refer to the value of $n$ in $Ni(CO)_n$. Reproduced with permission from De Kock, *Inorg. Chem.*, **10**, 1207 © 1971 American Chemical Society.

The formation of unsaturated metal carbonyls is of interest because similar products may be generated in low-temperature matrices by photolysis of stable binary metal carbonyl precursors (see Chapter 3). Thus $Cr(CO)_5$ may be produced by co-condensation of Cr atoms with excess CO at 4.2–10 K [2], or by ultraviolet photolysis of matrix-isolated $Cr(CO)_6$ [3], and it is interesting to compare the structure of $Cr(CO)_5$ molecules, formed by these different methods. At first it was thought that a $D_{3h}$ form, (1), of $Cr(CO)_5$ was produced by chromium atom reactions, which contrasted with the $C_{4v}$ geometry of $Cr(CO)_5$, (2), generated by photolysis. However, this claim to $D_{3h}$ $Cr(CO)_5$ was refuted [4] and later shown to be erroneous, and it appears that the structures of metal carbonyl fragments produced by metal atom co-condenation reactions are identical to those of the same molecules produced by photolysis.

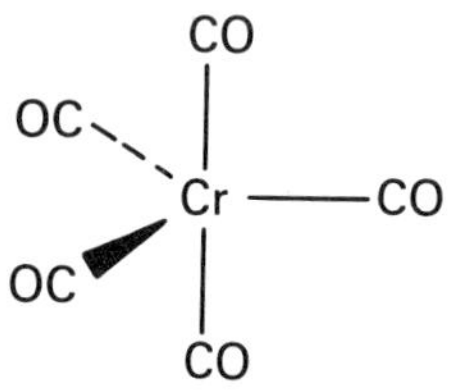

(1) 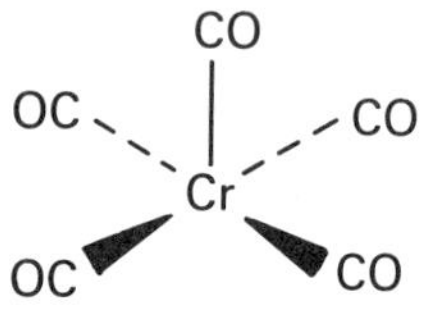                                                    (2)

The 17-electron metal carbonyl fragments $Mn(CO)_5$ [5] and $Co(CO)_4$ [6] may, likewise, be made by co-condensation of manganese or cobalt atoms with CO/Ar mixtures or pure CO. Manganese reacts with carbon monoxide, not only in its monatomic form to yield $Mn(CO)_5$, but also as a dimer to form two separate CO-bridged species each containing two manganese atoms. These complexes are perhaps best formulated as being $Mn_2(CO)$ and $Mn_2(CO)_2$, and the structures 3 and 4 have been tentatively assigned.

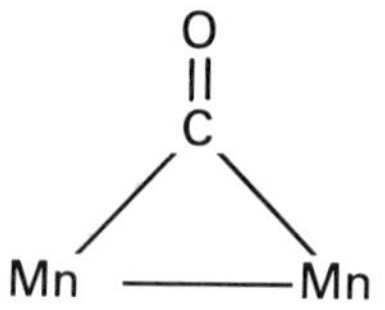

(3) 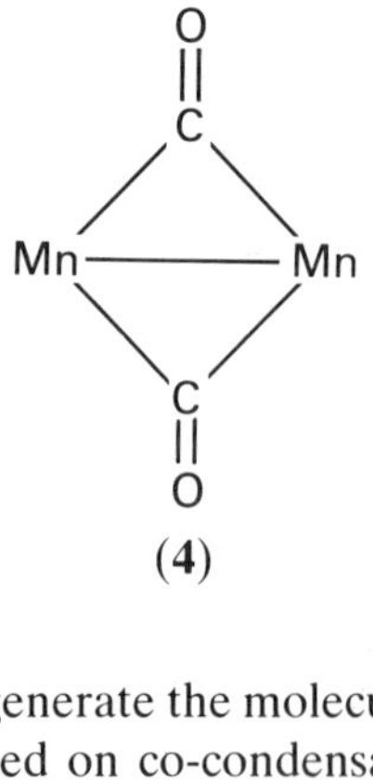                                                    (4)

A wide variety of methods have been used to generate the molecule $Co(CO)_4$ in low-temperature matrices. Not only is it produced on co-condensation of cobalt atoms with CO [6] but it can also be made by various photolytic routes. Thus

$Co(CO)_4$ is seen as a product when $Co(CO)_3(NO)$, $HCo(CO)_4$ or $Co_2(CO)_8$ is subjected to ultraviolet-visible irradiation in carbon monoxide matrices. Obviously the formation of a single product by a variety of routes helps considerably in its identification and characterization.

Although it is of interest to compare metal carbonyls formed by metal atom and photolytic reactions, the metal atom co-condensation technique also allows carbonyls to be formed from metals where there are no potential photochemical precursors. Thus metal carbonyl structure and bonding may be explored in new regions of the Periodic Table.

Originally it was thought that only metals in the three main transition series would form simple carbonyls. However, co-condensation of tin or aluminum atoms with mixtures of carbon monoxide and krypton at 20 K leads to the formation of tin [7] or aluminium [8] carbonyls. These products are readily detected by observation of C−O stretching vibrations in their infrared spectra. The most likely formulae for the major products of these reactions are $Sn(CO)$ and $Al(CO)_2$ respectively. Likewise uranium carbonyls are produced when a mixed matrix of carbon monoxide and argon containing uranium atoms is annealed [9]. As shown by the infrared spectra in Fig. 4.2 a series of molecules with the general formula $U(CO)_x$ ($x = 1$–6) is formed. When

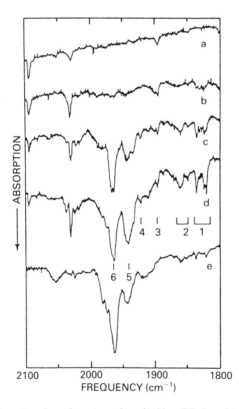

Fig. 4.2 — Infrared spectra of uranium atoms deposited in a CO-doped Ar matrix, followed by annealing: (a) after deposition; (b) 17 K; (c) 19 K; (d) 20 K; (e) 30 K. The numbers refer to the value of $n$ in $U(CO)_n$. Reproduced with permission from Slater *et al.*, *J. Chem. Phys.*, **55**, 5129 © 1971 American Institute of Physics.

the matrix is annealed to a temperature of 30 K the final binary carbonyl product appears to be $U(CO)_6$, probably with octahedral symmetry (5).

(5)                                                                                (6)

Lastly binary carbonyls of silver have attracted some interest. The first silver carbonyl was detected by Ogden from the infrared spectrum of a carbon monoxide matrix containing silver atoms [10]. More recently the products of the reaction between silver atoms and CO have been studied in some detail. Thus the hitherto unknown paramagnetic green complex, $Ag(CO)_3$ is seen, *inter alia*, as a product, and it has been characterized by infrared, ultraviolet-visible and electron spin resonance spectroscopy [11]. It appears to adopt a regular trigonal planar, $D_{3h}$, structure, **6**, in argon, krypton and xenon matrices, whereas a slight distortion of the structure is indicated in a carbon monoxide matrix. In a CO matrix, $Ag(CO)_3$ will dimerize to $Ag_2(CO)_6$ on annealing,

$$2Ag(CO)_3 \xrightarrow[30-37\,K]{CO\,matrix} Ag_2(CO)_6$$

and the kinetics of this reaction have been studied over the temperature range 30–37 K [12]. Thus the activation barrier to the dimerization process is calculated to be about $8 \, kJ \, mol^{-1}$.

## 4.3  FORMATION OF BINARY METAL OXIDES

As metal carbonyls may be produced by reactions of metal atoms with CO, so metal oxides may be formed when metal atoms react with $O_2$, $O_3$ or $N_2O$. In this way, several metal oxides have been trapped and characterized in low-temperature matrices.

Alkali metal oxides have proved to be fruitful subjects for study. Co-condensation of lithium atoms with $O_2/Ar$ mixtures at 16 K leads to the formation of $LiO_2$ [13]. This molecule has been characterized by Raman spectroscopy including isotopic substitution with $^6Li$ and $^{18}O$. The Raman spectra showing vibrations assigned to $v(O-O)$ and $v_{sym}(LiO_2)$ modes of different isotopomers of this molecule are given in Fig. 4.3. The most likely formulation of the molecule is an ion pair $Li^+O_2^-$, with side-on-co-ordination of the $O_2^-$ ion (7). In a similar manner, co-condensation of sodium atoms with $O_2/Ar$ mixtures at 15 K gives $NaO_2$, whose structure is analogous to **7**, and $Na_2O_2$ which has the $\mu$-peroxide structure, **8** [14].

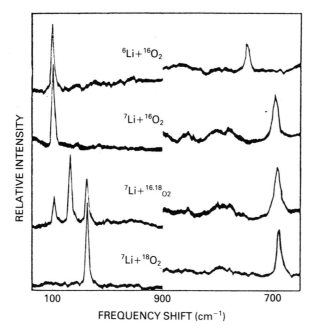

Fig. 4.3 — Raman spectra of the products obtained by co-condensation of Li atoms with 1:100 $O_2$/Ar mixtures. Reproduced with permission from Andrews and Smardzewski, *J. Chem. Phys.*, **58**, 2258 © 1973 American Institute of Physics.

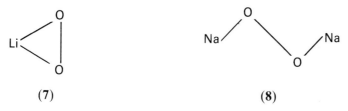

(7)                                                    (8)

If, on the other hand, lithium or sodium atoms are condensed with gas mixtures of argon and ozone, $O_3$, at 16 K, the ozonide species $MO_3$ (M = Li or Na) are produced [15]. An analogous product is obtained on reaction of caesium atoms with ozone, while co-deposition of caesium atoms with $O_2$/Ar mixtures at 15 K yields not only $CsO_2$ and $Cs_2O_2$ products analogous in structure to **7** and **8** but also a disuperoxide species $CsO_4$ [16]. Various structures may be envisaged for the $CsO_4$ molecule, but perhaps the most plausible candidate is a puckered five-membered ring (**9**). The wavenumbers of infrared and Raman bands assigned to the alkali metal peroxide, $MO_2$, molecules in argon matrices are listed in Table 4.1.

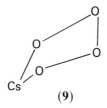

(9)

**Table 4.1** — Wavenumbers of infrared and Raman bands of alkali metal peroxide molecules in argon matrices

| Molecule | $v(O-O)$ | $v_{sym}(MO_2)$ | $v_{asym}(MO_2)$ |
|----------|----------|-----------------|------------------|
| $^6LiO_2$ | 1097.4 | 743.8 | 507.3 |
| $^7LiO_2$ | 1096.9 | 698.8 | 492.4 |
| $NaO_2$ | 1094 | 390.7 | 332.8 |
| $KO_2$ | 1108 | 307.5 | — |
| $RbO_2$ | 1111.3 | 255.0 | 282.5 |
| $CsO_2$ | 1115.6 | 236.5 | 268.6 |

Other main group oxide molecules have been produced by metal atom reactions and studied in low-temperature matrices. An example of such a reaction is provided by that between tin atoms and $O_2$ to form $SnO_2$ [17]. In this study, the linear $SnO_2$ molecule (**10**) was seen to undergo a further reaction with tin atoms to form the cyclic $Sn_2O_2$ molecule (**11**). Small yields of $SnO$, $SnO_3$ and $O_3$, implying the formation of

$$O=Sn=O \quad + \quad Sn\diagup \longrightarrow \quad \begin{matrix} & O & \\ Sn & & Sn \\ & O & \end{matrix}$$

(**10**)                                                                                    (**11**)

atomic oxygen, were also seen. Fig. 4.4 illustrates the infrared spectra obtained in this experiment.

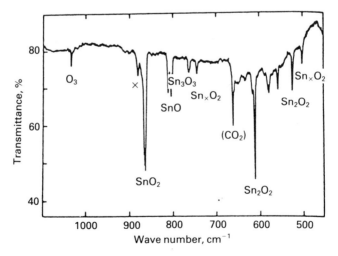

Fig. 4.4 — Infrared spectrum obtained after condensing tin vapour with Kr doped with 15% $O_2$. Reproduced with permission from Bos and Ogden, *J. Phys. Chem.*, **77**, 1514 © 1973 American Chemical Society.

Similar experiments have been carried out with transition metal atoms, and a wide array of metal oxide structures have been observed. Chromium atoms react on co-condensation with $O_2$ to form both the non-linear dioxo molecule, $CrO_2$ (**12**) [18a] (which is the same product as is observed when $Cr(CO)_6$ is subjected to prolonged ultraviolet irradiation in $O_2$-containing matrices — see section 3.9), and the bis-peroxide molecule $Cr(O_2)_2$ which probably has the structure **13** [18b]. Copper atoms also react under similar conditions, with $O_2$ to form a $Cu(O_2)_2$ product, with structure **13** [18b]. However, photolysis of $O_2$-doped matrices containing either manganese or copper atoms leads to the formation of the monoxide molecules MnO and CuO [19]. Although condensation of nickel atoms with dioxygen or $O_2$/Ar mixture yields $NiO_2$ [20], this molecule does not have the non-linear dioxo structure (**12**) shown by $CrO_2$, but rather a triangular peroxo structure (**14**). However, when a matrix containing nickel atoms and $O_2$ molecules is subjected to annealing, it appears that the product is again a bis-dioxygen complex, $Ni(O_2)_2$ (**13**), analogous to both $Cr(O_2)_2$ and $Cu(O_2)_2$. It is interesting to note that while the $MO_4$ complexes of chromium, copper and nickel adopt similar structures, this structure is quite different from that proposed for the oxide $CsO_4$ described earlier.

Finally co-condensation of uranium atoms with $O_2$/Ar mixtures yields, *inter alia*, the binary uranium oxides $UO_2$ and $UO_3$ [21]. $UO_2$, like $CrO_2$, is a non-linear dioxo molecule (**12**) while $UO_3$ exhibits the interesting T-shaped structure (**15**). These same uranium oxides have also been trapped in low-temperature matrices, by heating a sample of solid $UO_2$ in a Knudsen cell to 2175 K, and condensing the vapour over the sample with excess argon. In both experiments the assignments of bands are in agreement.

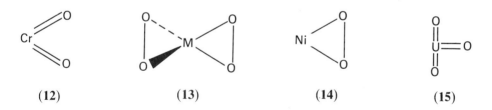

(**12**)                    (**13**)                    (**14**)                    (**15**)

The reactions of metal atoms with mixtures of $O_2$ and CO have received some attention. However, the products of these reactions, which bear a resemblance to the photochemical intermediates described in section 3.9, tend to have complicated structures and are difficult to characterize unambiguously. Co-condensation of gold atoms with an equimolar mixture of CO and $O_2$ at 10 K yields a single complex which is perhaps best formulated as a monocarbonyl gold(II) peroxoformate (**16**) [22]. By contrast, cobalt atoms react on co-condensation with $CO/O_2$ mixtures to give a number of products with the general formula $Co(O_2)(CO)_n$, where $n = 1-4$ and $O_2$ is co-ordinated in either a side-on or an end-on manner [23]. It is proposed that the product $Co(CO)_4(O_2)$ is best described as superoxide ion pair (**17**), where the $[Co(CO)_4]^+$ component of the pair adopts a distorted tetrahedral structure, analogous to the structure seen for its isoelectronic neighbour $Fe(CO)_4$. The structural assignments of these products remain, however, tentative.

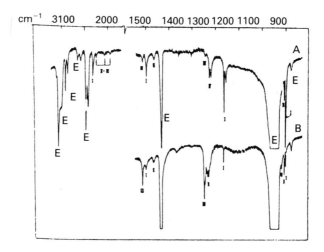

(16)                                        (17)

## 4.4  FORMATION OF ALKENE AND ALKYNE COMPLEXES

Alkene and alkyne complexes of various main group and transition metals have been prepared by co-condensation of metal atoms with gas mixtures containing reactive unsaturated hydrocarbon molecules. The trapped complexes may then be studied spectroscopically. A particular interest in these experiments is to determine the mode of co-ordination of the alkene or alkyne ligand to the metal centre.

An efficient route to binary nickel ethylene complexes of general formula $Ni(C_2H_4)_x$ ($x = 1, 2$ or 3). is co-condensation of nickel atoms with $C_2H_4$ or $C_2H_4/Ar$ mixtures at 15 K [24]. The products have been identified by infrared and ultraviolet-visible spectroscopy and Fig. 4.5 illustrates the infrared spectrum observed for an argon matrix containing Ni atoms, and $C_2H_4$. The increase in concentration of the product containing three ethylene ligands, when the matrix is annealed to 35 K, can clearly be seen. Table 4.2 lists the infrared absorptions observed for these complexes.

Electron spin resonance spectroscopy has also been used to study paramagnetic metal alkene complexes. An advantage of this approach is that information may be

Fig. 4.5 — Infrared spectra obtained after condensing Ni atoms with a 1:50 $C_2H_4/Ar$ mixture: (a) after deposition; (b) after annealing to 35 K. The bands are assigned as follows: E = free $C_2H_4$; I = $(C_2H_4)Ni$; II = $(C_2H_4)_2Ni$; III = $(C_2H_4)_3Ni$. Reproduced with permission from Huber *et al.*, *J. Am. Chem. Soc.*, **98**, 6508 © 1976 American Chemical Society.

**Table 4.2** — The infrared absorptions ($v$/cm$^{-1}$) of Ni(C$_2$H$_4$)$_x$ ($x$ = 1, 2 or 3)

| (C$_2$H$_4$)$_3$Ni | (C$_2$H$_4$)$_2$Ni | (C$_2$H$_4$)$_2$Ni | Assignment |
|---|---|---|---|
| 2918 | 2945 | ~2960 | $v$(C−H) |
|  | 2886 |  |  |
| 1512 | 1465 | 1496 | $v$(C=C) |
| 1245 | 1235 | 1158 | $\delta$(CH$_2$) |
|  | 1244 |  |  |
| — | 906 | ~900 | $\rho$(CH$_2$) |

obtained about the 'spin density' within the molecule, and thus questions about the exact mode of bonding of the alkene ligand to the metal may be answered. Fig. 4.6 illustrates the ESR spectrum obtained when aluminium atoms are co-condensed with

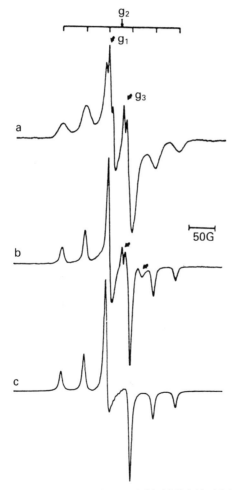

Fig. 4.6 — ESR spectra of (a) (C$_2$H$_4$)Al and (b) (C$_2$D$_4$)Al; (c) is a computer-simulated spectrum. Reproduced with permission from Kasai and McLeod, *J. Am. Chem. Soc.*, **97**, 5609 © 1975 American Chemical Society.

$C_2H_4$/Ne and $C_2D_4$/Ne mixtures [25]. A prominent sextet feature is seen and this is attributed to hyperfine interaction with the $^{27}Al$ nucleus which has 100% natural abundance and a value of $l = 5/2$. In principle, three bonding schemes are possible for

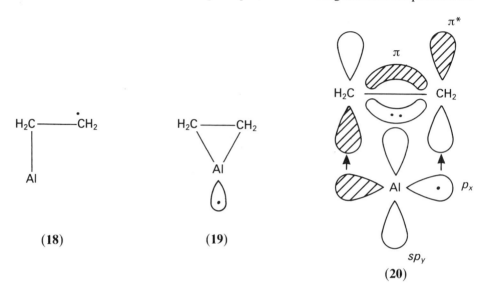

**(18)**                **(19)**                **(20)**

the complex $Al(C_2H_4)$ (**18**, **19** and **20**). Model **19** invokes $sp^2$ hybridization of the Al atom and formation of two Al–C σ bonds. Model **20** involves an $s-p_y$ hybridization of Al, the donation of electrons from the π orbital of $C_2H_4$ into the vacant Al $sp_y$ hybrid and back-donation from the semi-filled $p_x$ orbital of Al into the $C_2H_4$ $\pi^*$ antibonding orbital. From the ESR measurements it is found that the unpaired eletron resides in an orbital of mainly $p$-type character since the spin densities in the $3s$ and $3p$ Al orbitals in the complex are found to be 0.015 and 0.70 respectively. Thus, it appears that model **20** more closely represents the true structure of $Al(C_2H_4)$.

Similar ESR experiments have been carried out on the product formed when aluminium atoms are co-condensed with gas mixtures containing acetylene [26]. In this case a π-bonded complex is not observed, but rather the principal product is an adduct with the vinyl structure (**21**). The ESR spectrum of the perdeuterated form of this adduct $Al(C_2D_2)$ is shown in Fig. 4.7 where the sextet structure, resulting from hyperfine interaction of the unpaired electron to the $^{27}Al$ nucleus, can clearly be seen.

Recent experiments have shown that iron atoms will interact with ethylene molecules in low-temperature matrices in two distinct ways [27]. First two series of

$$\begin{array}{c} \text{H}\diagdown \\ \phantom{xx} \diagup\text{C}=\dot{\text{C}}\text{H} \\ \text{Al} \end{array}$$

**(21)**

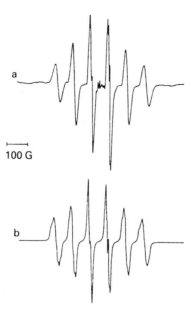

Fig. 4.7 — ESR spectrum observed from a matrix containing Al atoms and $C_2D_2$; (b) is a computer-simulated spectrum. Reproduced with permission from Kasai *et al.*, *J. Am. Chem. Soc. Soc.*, **99**, 3521 **112** 1966 American Chemical Society.

classically $\pi$-bonded complexes $Fe(C_2H_4)_x$ ($x \geqslant 2$) and $Fe_2(C_2H_4)_y$ ($y = 1$ or 2) are produced. In addition to these complexes, however, an unusual H-bonded product is obtained. There are two forms of this product and each exhibits a slightly perturbed ethylene-like infrared spectrum. Similarly an H-bonded product is obtained when Fe atoms are co-condensed with acetylene [27]. The structures **22**, **23** and **24** can be postulated for these complexes.

    (22)                (23)                (24)

In this way some idea of the different modes of alkene or alkyne bonding to metal centres is beginning to emerge. Such information is of importance because these matrix-isolated complexes may bear structural similarities to intermediates in reactions of unsaturated hydrocarbons, for example, alkene hydrogenation and dimerization processes which are catalysed by transition metal species.

## 4.5   C−H and C−C BOND ACTIVATION

The hydrogen-bonded iron complexes described in section 4.4 are of particular interest because they provide examples of molecules which will undergo oxidative-addition reactions of the type

$$Fe(C_2H_4) \underset{\lambda \geqslant 400\,nm}{\overset{\lambda = 280-360\,nm}{\rightleftharpoons}} HFe(C_2H_3)$$

and

$$Fe(C_2H_2) \xrightarrow[\text{irradiation}]{\text{near UV-visible}} HFeC_2H$$

In the case of the molecule $HFe(C_2H_3)$, the reaction is photochromic, since irradiation with visible light of wavelengths greater than 400 nm regenerates the H-bonded $C_2H_4$ complex of iron. Reactions of this sort, involving insertion into C−H bonds, evoke great interest because of the possibility that they may be used to 'activate' inert hydrocarbons and therefore open up new synthetic possibilities.

The most simple hydrocarbon is methane, $CH_4$, and it has been found that photo-excited iron atoms will insert into a C−H bond of this molecule at 10–12 K. This reaction is not undergone by ground-state iron atoms, but is only seen when iron atoms, trapped in a methane matrix, are subjected to irradiation at 300 nm. A similar photoinsertion reaction has been observed for manganese, cobalt, copper, zinc, silver and gold atoms trapped in methane matrices at 10 K [28a]. Table 4.3 lists the infrared bands of these $CH_3MH$ products.

**Table 4.3** — Vibrational wavenumbers (cm$^{-1}$) of matrix-isolated $CH_3MH$ molecules

| Metal | $\upsilon$(C−H) | $\upsilon$(M−H) | $\delta,\rho$(CH$_3$) | | $\upsilon$(M−C) |
|-------|------|------|------|------|------|
| Mn | 2932.9 | 1582.9 | 1142.3 | 550.3 | 500.7 |
| Fe | 2933.5 | 1653.1 | 1153.4 | 548.8 | 521.1 |
|    |        |        |        | 545.9 |      |
| Co | —      | 1699.5 |        | 585.4 | 527.5 |
|    |        |        |        | 576.7 |      |
| Cu | —      | 1855.7 | 1200.1 | 613.8 | 433.9 |
| Ag | 2907.6 | 1725.8 | 1232.4 | 614.7 | —    |
|    | 2900.1 |        |        |       |      |
| Au | —      | 2195.8 | 1202.8 | 610.9 | —    |
| Zn | —      | 1845.8 | 1069.5 | 689.1 | 447.1 |

The observation of a single M−H and a single M−C stretching mode for each complex argues in favour of a methyl metal hydride structure (**25**) rather than the alternative methylene metal dihydride insertion product (**26**). Moreover, the M−H stretching mode occurs at a frequency which is normal for such vibrations. This observation, in turn, argues against a third plausible alternative structure (**27**) where the M–H bond would be expected to be considerably weakened. Interestingly, as with the insertion product of iron atoms and ethylene, reductive elimination

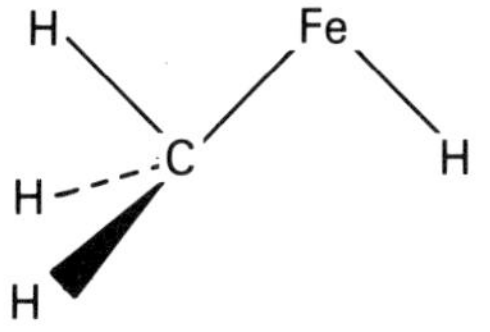

(25) 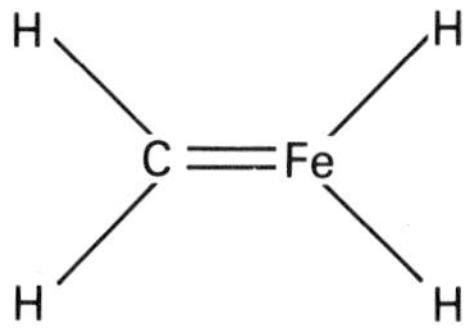                    (26) 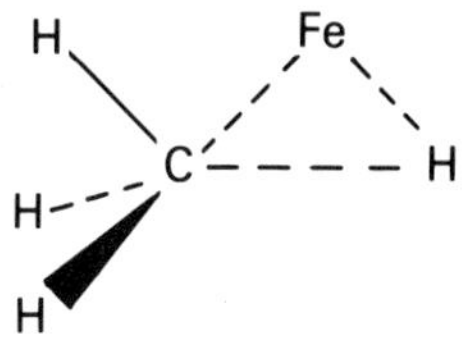                    (27)

is caused by photolysis at longer wavelengths [28b]. Thus irradiation of the $CH_3FeH$ molecule at 420 nm causes the regeneration of Fe atoms.

$$CH_3FeH \underset{300\,nm}{\overset{420\,nm}{\rightleftharpoons}} Fe + CH_4$$

A similar C–H insertion reaction will occur with ethane, whereas iron atoms insert preferentially into the C–C bond of cyclopropane to form the four-membered metallocyclic molecule (28) [29].

$$C_2H_6 \quad + \quad Fe \xrightarrow{\;UV\;} C_2H_5FeH$$

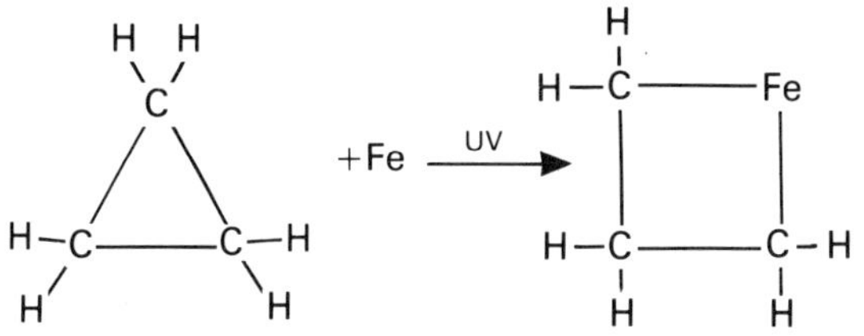

(28)

These metal atom insertion reactions complement other studies in which photochemically generated species will activate C–H bonds. It has been found that when the stable 18-electron molecule $Cp_2WH_2$ (29) $(Cp = \eta\text{-}C_5H_5)$ is subjected to photolysis in benzene solutions, the product $Cp_2W(C_6H_5)H$ is formed, in which the tungsten metal centre has inserted into a C–H bond of benzene. It is believed that the key intermediate in this reaction is the unstable 16-electron metallocene $Cp_2W$ (30). This product has not been isolated in solution, but has been produced by

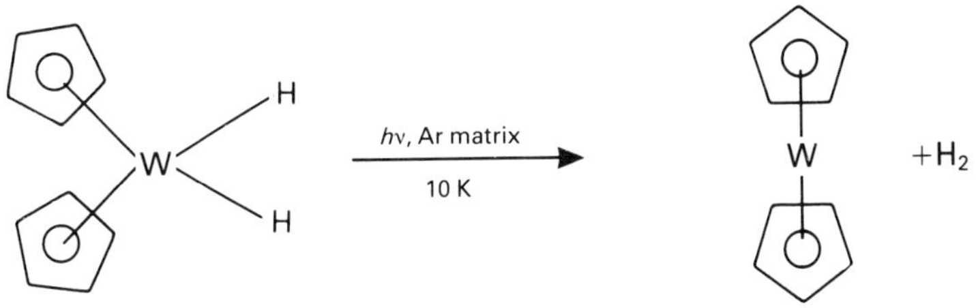

(29)                                                          (30)

photolysis of $Cp_2WH_2$ in an argon matrix at 10 K [30a]. More recently it has been shown that many unsaturated organometallic complexes of rhodium or iridium produced, for example, by photolysis of alkene complexes, are potent activators of C−H bonds. An example is provided by the rhodium complex **31** [30b].

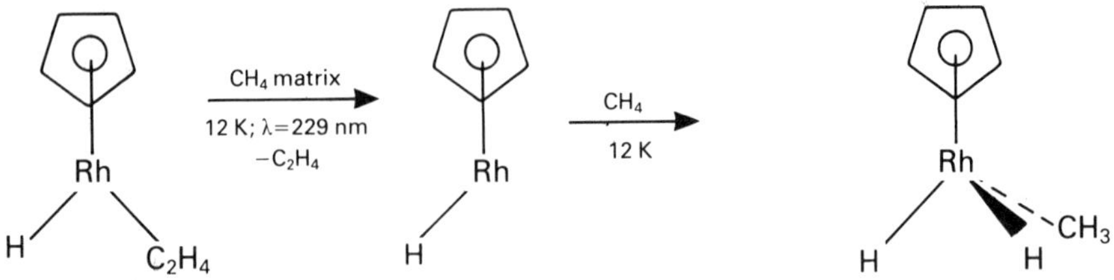

**31**

## 4.6 OTHER INSERTION REACTIONS

Insertion reactions of matrix-isolated metal atoms are not limited to C−H and C−C bonds. Many different metal atoms are found to insert into the O−H bond of water, to form HMOH molecules [31]. Scandium, titanium and vanadium atoms will insert spontaneously when the metal atoms are co-condensed with $H_2O$/Ar mixtures at 15 K. Chromium, manganese, iron, cobalt, copper and zinc atoms all form metal–water adducts on deposition, while ultraviolet irradiation of these adducts leads, with the exception of the zinc adduct, to O−H bond insertion and formation of HMOH products. In the case of zinc, the adduct preferentially reacts to form ZnOH, while CuOH is observed, as the principal product, alongside HCuOH, from the copper adduct. Only nickel of the first row transition metals fails to yield either an adduct or an insertion product under these conditions. The reactions may be summarized as follows:

$$M + H_2O \xrightarrow[15\,K]{Ar\,matrix} HMOH \quad (M = Sc, Ti\ or\ V)$$

$$M' + H_2O \xrightarrow[15\,K]{Ar\,matrix} M'(OH_2) \xrightarrow[irradiation]{UV} HM'OH \quad (M' = Cr, Mn, Fe, Co\ or\ Cu)$$

$$M'' + H_2O \xrightarrow[15\,K]{Ar\,matrix} M''(OH_2) \xrightarrow[irradiation]{UV} M''OH \quad (M'' = Cu\ or\ Zn)$$

Table 4.4 lists some estimated enthalpies for the reactions $M + H_2O \rightarrow HMOH$. These processes are somewhat complicated by the insertion of $M_2$ dimers into O−H bonds to form $HM_2OH$ and HMOMH products. However, the initial $M(OH_2)$ adducts are characterized by the bending mode ($\nu_2$) of the co-ordinated water molecule, which is shifted to low frequency by some 5–30 cm$^{-1}$ with respect to free, unco-ordinated water. The HMOH products, on the other hand, are characterized

**Table 4.4** — Estimated heats of reaction for the processes $M + H_2O \rightarrow HMOH$

| Metal | $\Delta H^0_{rxn}$ | Comments |
|-------|------|----------|
| Sc | − 230 | Reaction occurs on deposition |
| Ti | − 209 | Reaction occurs on deposition |
| V  | − 197 | Reaction occurs on deposition |
| Cr | − 92 | Reaction occurs on photolysis |
| Mn | − 105 | Reaction occurs on photolysis |
| Fe | − 117 | Reaction occurs on photolysis |
| Co | − 92 | Reaction occurs on photolysis |
| Ni | − 88 | No products formed |
| Cu | − 13 | Minor reaction pathway only |
| Zn | − 21 | Different reaction pathway followed |

by $\nu(M-H)$ and $\nu(M-O)$ modes. Table 4.5 lists the wavenumbers of some of these absorptions, and also gives calculated force constants for the $M-H$ and $M-O$ bonds of the HMOH products.

Photoexcited iron atoms will insert into either the $O-H$ or the $C-O$ bond of methanol in an argon matrix at 14 K to form the products $CH_3OFeH$ and $CH_3FeOH$ [32]. Initially, deposition of iron atoms with a mixture of methanol and argon at 14 K yields the adduct $Fe(CH_3OH)$. Photolysis of this adduct with visible light of wavelengths longer than 400 nm causes the Fe atom to insert into the $O-H$ bond of methanol to give the product $CH_3OFeH$, characterized by an $Fe-H$ stretching vibration at $1741.0 \, cm^{-1}$. However, if the matrix is then subjected to irradiation with ultraviolet light, at wavelengths between 280 and 360 nm, the concentration of $CH_3OFeH$ decreases, while that of a new product, $CH_3FeOH$, characterized by an $O-H$ stretching vibration at $3744.8 \, cm^{-1}$ and an $Fe-O$ stretching vibration at $687.5 \, cm^{-1}$, increases. These changes are illustrated by the infrared spectra in Fig. 4.8 and are summarized below.

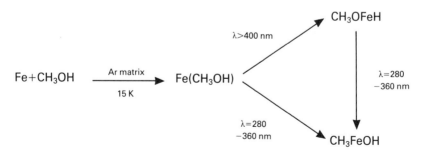

A further interesting study has shown that the methylene complex, $Fe=CH_2$, is generated by co-condensation of iron atoms with diazomethane, $CH_2H_2$, and excess argon at 14 K [33]. Simple alkylidene complexes of this sort are of great interest, because they are now recognized as being similar to major intermediates in the important alkene metathesis process.

**Table 4.5** — Wavenumbers (cm$^{-1}$) of infrared bands of M(OH$_2$) adducts and HMOH products

| Metal | M(OH$_2$) adduct | | HMOH product | | | |
| | $\nu_2$(H$_2$O)/cm$^{-1}$ | $\Delta\nu_2$(H$_2$O)/cm$^{-1}$ | M–H | | M–O | |
| | | | $\nu$/cm$^{-1}$ | $k$/mdyn Å$^{-1}$ | $\nu$/cm$^{-1}$ | $k$/mdyn Å$^{-1}$ |
|---|---|---|---|---|---|---|
| Sc | — | — | 1485.1 | 1.34 | 715.8 | 3.73 |
| Ti | — | — | 1538.9 | 1.44 | 699.7 | 3.79 |
| V | — | — | 1583.0 | 1.54 | — | — |
| Cr | 1580.2 | 13.1 | 1639.9 | 1.75 | 674.1 | 3.40 |
| Mn | 1576.8 | 16.5 | 1663.4 | 1.76 | 648.1 | 3.18 |
| Fe | 1562.8 | 30.5 | 1731.9 | 1.90 | 682.4 | 3.53 |
| Co | 1564.3 | 29.0 | 1790.4 | 2.08 | 667.4 | 3.37 |
| Ni | — | — | — | — | — | — |
| Cu | 1572.8 | 20.5 | 1910.8 | 2.40 | 615.6 | 3.13 |
| Zn | 1587.7 | 5.6 | — | — | — | — |
| (Water) | (1593.3) | (0.0) | — | — | — | — |

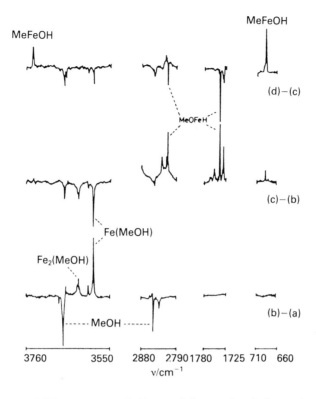

Fig. 4.8 — Infrared difference spectra of adducts and photoproducts in the reaction of Fe atoms with methanol in solid argon. Reproduced with permission from Park *et al.*, *J. Chem. Soc.*, *Chem. Commun.*, 1570 © 1985 Royal Society of Chemistry.

## 4.7  FORMATION OF FREE RADICALS

Reactions of metal atoms in matrices may be used to generate free radicals. This approach was pioneered by Andrews and Pimentel [34] who succeeded in producing methyl radicals in argon matrices by co-condensing lithium atoms with methyl halides and excess argon at 15 K. The reactions may be summarized by the equation,

$$CH_3X + Li \longrightarrow CH_3 + LiX \ ,$$

where X is a halogen atom. The methyl radicals so produced were characterized by infrared spectroscopy, including isotopic substitution with $^{13}C$ and $^2H$. On annealing, infrared absorptions due to ethane appear and grow at the expense of those assigned to the methyl radical

$$2CH_3 \xrightarrow[\text{anneal}]{\text{Ar matrix}} C_2H_6$$

These experiments are of interest because they were the first to demonstrate the formation of matrix-isolated methyl radicals. Other approaches to generating methyl radicals in matrices had been attempted, but all had been frustrated. For example, gas phase pyrolysis of dimethyl mercury, diazomethane or di-*tert*-butyl peroxide, followed by trapping with argon or freons, produced only stable products, while *in situ* photolysis of diazomethane, nitromethane or methyl iodide, in solid argon, was unsuccessful in each case, undoubtedly as a result of the matrix cage effect.

However, other routes to matrix-isolated methyl radicals are now known. Milligan and Jacox have produced methyl radicals by vacuum-ultraviolet photolysis of methane in argon or nitrogen matrices, Snelson has succeeded in obtaining the same product by gas phase pyroloysis of dimethyl mercury or methyl iodide, followed by trapping of the products with excess neon [35]. The results of these later experiments suggest that some of the infrared absorptions observed by Andrews and Pimentel should be assigned to a new methyl lithium halide ($CH_3LiX$) molecule, rather than to the methyl radical itself. Burdett, meanwhile, has rationalized the vibrational spectra of free and alkali halide-perturbed methyl radicals in low-temperature matrices, in terms of an interaction between the alkali metal halide and the lowest-lying unoccupied, antibonding orbital of the methyl radical [36]. These experiments demonstrate the difficulty of obtaining a reactive species, which is truly isolated within a matrix. However, they also show that the interactions between different species in a matrix may be understood, if one of the species in a matrix may be generated from a variety of different sources.

## 4.8  STUDIES OF METAL DIMERS

When a metal vapour is condensed with an excess of inert gas, the condensate is often found to contain not only metal atoms, but also small molecules containing two or more of these atoms. The concentration of dimers and small clusters may be increased by adjusting the experimental conditions of deposition, for example, by changing the matrix material, increasing the concentration of the metal, increasing the temperature at which the substrate is hcld during deposition, or by annealing the matrix after deposition. In some cases photolysis is found to lead to changes in the relative concentrations of the different species.

Matrix-isolated metal dimers have been investigated by a variety of techniques. Optical absorption or emission spectroscopy has been the principal method, but Raman, resonance Raman and electron spin resonance spectroscopy have all been used in many experiments. Recently, EXAFS studies have been made on metal dimers and small clusters. Studies of diatomic metal molecules are of interest because such molecules represent the first stage in the nucleation of a metal in the gas phase. However, many of the molecules are of interest in their own right. For example, the strength of bonding of a metal dimer varies enormously from the weakly bound ground states of $Mg_2$ and $Ca_2$, which are essentially van der Waals molecules, to the molecule $Mo_2$ which is extremely strongly bound in its ground state. From the large number of experiments that have been carried out, two areas of study will served to exemplify the interesting results that have been obtained for matrix-isolated metal dimers.

First, the chromium dimer, $Cr_2$, presents an interesting case history. This molecule has been produced by co-condensation of chromium vapour with excess argon, and it has been detected by resonance Raman spectroscopy [37]. Fig. 4.9 shows the lines in the resonance Raman spectrum, which are assigned to $Cr_2$,

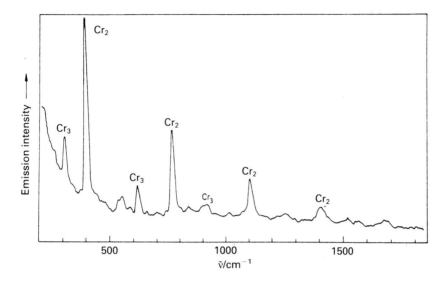

Fig. 4.9 — Resonance Raman spectrum of $Cr_2$ and $Cr_3$ in an argon matrix. Reproduced with permission from DiLella *et al.*, *J. Chem. Phys.*, **77**, 5263 © 1982 American Institute of Physics.

alongside those lines which are assigned to the $C_{2v}$ trimer, $Cr_3$. From the vibrational progression, it is possible to calculate a ground-state harmonic frequency of $427.5\,\text{cm}^{-1}$ and a force constant for the Cr−Cr bond of $2.80\,\text{mdyn}\,\text{Å}^{-1}$. This indicates that $Cr_2$ is rather more weakly bound than $Mo_2$, in its ground state, since $Mo_2$ has an enormous ground-state force constant for the Mo−Mo bond of $6.44\,\text{mdyn}\,\text{Å}^{-1}$.

An interesting observation has been made on matrix-isolated $Cr_2$ molecules which are pumped with radiation corresponding to the $x'\Sigma_g \rightarrow A'\Sigma_u$ absorption ($\lambda = 458\,\text{nm}$) of the molecule. A new absorption is seen under these conditions which appears to belong to a metastable state of $Cr_2$, whose lifetime ranges from 0.2 to 0.7 s depending on the matrix material employed [38]. This new state has vibrational constants of $\omega_e = 78.6$ and $\omega_e x_e = 0.4\,\text{cm}^{-1}$, and the most likely explanation of these results is that the new species is a long-bonded form of $Cr_2$ (with a correspondingly weakened Cr−Cr bond) trapped in the outer minimum of a double minimum ground-state potential. Fig. 4.10 gives a proposed scheme by which the outer minimum of such a potential energy surface may be populated.

A second interesting story is provided by the dimers of bismuth and lead [39], $Bi_2$ and $Pb_2$. Either of these molecules may be generated by condensation of the appropriate metal vapour with an excess of an inert gas. An interesting aspect of these studies is the detection of Raman scattering within electronically excited states of the molecules. Fig. 4.11 illustrates the Raman spectrum of the $Bi_2$ in its ground ($x$) state and an electronically excited $Bi_2^\star$ state. From this spectrum it is possible to calculate vibrational constants of $\omega_e = 173.5$ and $\omega_e x_e = 0.37\,\text{cm}^{-1}$ for the ground state and of $\omega = 133.3$ and $\omega_e x_e = 0.31\,\text{cm}^{-1}$ for the excited state of $Bi_2$. It remains unclear, however, exactly which excited state of $Bi_2$ is responsible for the Raman scattering. In both $Pb_2$ and $Bi_2$ Raman scattering is observed from excited states as a

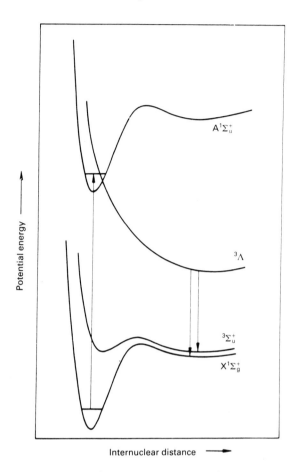

Fig. 4.10 — Proposed scheme for the population of the outer minimum of the double-minimum potential curve believed to apply to ground state $Cr_2$. Reproduced with permission from Moskovits *et al.*, *J. Chem. Phys.*, **82**, 4875 © 1985 American Institute of Physics.

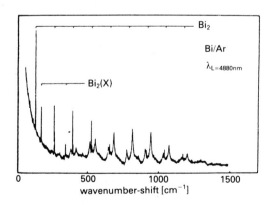

Fig. 4.11 — Raman spectrum of $Bi_2$ in its ground (X) and an electronically excited (*) state. Reproduced with permission from Eberle *et al.*, *Chem. Phys.*, **92**, 421 © 1985 Elsevier.

result of multiphoton processes, in which laser irradiation is not only responsible for exciting Raman scattering, but also excites the diatomic molecules from their ground to excited states.

A wide range of metal dimers have been trapped and studied in low-temperature matrices, and Table 4.6 summarizes some of the findings for diatomic molecules of the first-row transition metals.

**Table 4.6** — Ground-state harmonic frequencies and corresponding force constants for dimers of the first row transition, and post-transition metals

| Molecule | $\upsilon/cm^{-1}$ | $f/mdyn\ Å^{-1}$ |
|:---:|:---:|:---:|
| $Sc_2$ | 238.9 | 0.76 |
| $Ti_2$ | 407.9 | 2.35 |
| $V_2$ | 537.5 | 4.34 |
| $Cr_2$ | 427.5 | 2.80 |
| $Mn_2$ | 124.0 | 0.25 |
| $Fe_2$ | 300.26 | 1.48 |
| $Co_2$ | 290 | 1.46 |
| $Ni_2$ | 380.9 | 2.48 |
| $Cu_2$ | 264.55 | 1.30 |
| $Zn_2$ | 80 | 0.12 |

## 4.9  SMALL METAL CLUSTERS

Co-condensation of a metal vapour with an excess of an inert gas yields, in addition to metal dimers, small metallic cluster molecules. Various spectroscopic techniques have been used to study these matrix-isolated clusters, and for those with unpaired electrons the ESR technique has proved to be of particular importance. Typical of these matrix ESR studies is the work of Lindsay and his co-workers at the University of New York on small clusters of the alkali metals.

When lithium vapour is condensed with excess argon or nitrogen, a number of lithium cluster molecules are produced, including $Li_3$, and several of these have been structurally characterized by ESR spectroscopy [40]. The spectrum of the isotopomer $^6Li_3$ shows seven equally spaced transitions with relative intensities which are in good agreement with the expected values for a molecule with three equivalent nuclei with values of $I = 1$. This result is interpreted in terms of a rapidly pseudorotating trimer. Fig. 4.12 shows the ESR spectrum of $^6Li_3$ trapped in an argon matrix at 28.5 K. By contrast, the trimeric clusters, $Na_3$ and $K_3$, may be trapped in non-pseudorotating forms under matrix conditions. The ground state for both these molecules appears to have $C_{2v}$ symmetry where the apical angle is greater than 60° (**32**). This geometry is produced by Jahn–Teller distortion of an equilibrium triangular structure (**33**) of $D_{3h}$ symmetry. Larger clusters, too, may be produced. For example, $Li_7$ has been formed under similar conditions; it has two sets of

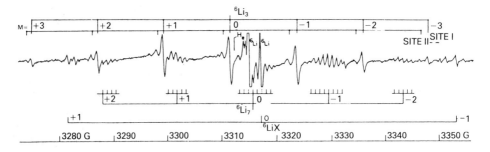

Fig. 4.12 — ESR spectrum of pseudorotating $^6Li_3$ in an argon matrix at 28.5 K. Reproduced with permission from Garland and Lindsay, *J. Chem. Phys.*, **78**, 2814 © American Institute of Physics.

(32)          (33)

equivalent nuclei: one set consisting of five lithium atoms and one of two lithium atoms, thereby implying the pentagonal bipyramidal structure **34** for this cluster.

The molecule $Ag_3$ has been trapped in a glass of $C_6D_6$ at 103 K and likewise, has been studied by ESR spectroscopy [41]. These studies suggest that $Ag_3$, like the

• = Li

(34)          (35)

alkali metal trimers, has a $C_{2v}$ non-linear ground state. Similarly both $Cu_3$ and CuAgCu appear to adopt this obtuse-angled structure when trapped in $C_6D_6$ glasses at ca. 100 K. Table 4.7 lists the findings from ESR experiments on $Ag_3$, $Cu_3$ and CuAgCu.. In this table, $\rho_M$ is the unpaired spin population on the metal atom M and it can be seen that in each of these trimers, most of the spin density resides on the two terminal atoms. The silver cluster $AG_5$ has also been trapped in $C_6D_{12}$ or adamantane glasses at 77 K [41]. This molecule is prepared by continuous irradiation at wavelengths greater than 320 nm of the silver-containing glass during deposition. It is found to have a distorted trigonal bipyramidal structure of $C_{2v}$ symmetry (**35**), and most of the spin density resides on the two equivalent axial silver nuclei.

**Table 4.7** — Spin populations of metal atoms of $M_3$ clusters

| Trimer | $g$ factor | $\rho_M$ (terminal) | $\rho_M$ (central) |
|--------|-----------|---------------------|--------------------|
| $Ag_3$ | 1.9622 | 0.44 | 0.06 |
| $Cu_3$ | 1.9925 | 0.29 | 0.026 |
| CuAgCu | 1.9621 | 0.41 | 0.054 |

An important recent development has been the use of the EXAFS technique to look at matrix-isolated metal clusters. This approach gives bond length information, which is not readily forthcoming from other spectroscopic methods used to study matrix-isolated species. The deposition of iron into inert gas matrices has been studied as a function of iron concentrations from 0.1 to 1.5 atom per cent Fe [42]. At the lowest concentration, only atoms and dimers are detectable and an Fe–Fe bond length of 1.87 Å is reported. This is a marked contraction from the 2.48 Å Fe−Fe distance in bulk iron metal. At 0.4–0.5 atom per cent Fe, trimers and tetramers are also present and the contraction is reduced; the mean Fe−Fe bond length becomes 2.02 Å. Larger particles occur at a concentration of 1.5 atom per cent Fe, but there is still a shortening of the Fe−Fe distance to 2.30 Å. Studies of this type are of importance because they give an insight into the mechanism by which metal atoms aggregate to yield a bulk metal.

## 4.10 REACTIONS OF METAL CLUSTERS

Only a few reactions of metal clusters have been studied in low-temperature matrices. One example, which has been reported, is the interaction of small copper clusters with CO [43]. In these experiments, species of the type $Cu_nCO$, where $n = 1$–4, are produced and it is found that the CO stretching frequency, for CO attached to the larger clusters ($n = 3$ or 4), is similar to that of CO adsorbed on polycrystalline copper.

A more successful approach to the investigation of the reactivity of metal clusters has been to study the reactions of such molecules in the gas phase in supersonic beams. Here a portion of a target rod of the metal under investigation is vaporized by the action of a pulsed laser beam and the metal vapour is entrained in a stream of helium, where nucleation and cluster growth occur rapidly. A reactant gas may be added to the helium and the mixture is then expanded into a vacuum through a small nozzle of ca. 1mm diameter. Gas phase reactions are terminated by this expansion process, and after collimation of the beam, products may be detected by mass spectrometry. In this way, the reaction of hydrogen with various transition metal clusters has been studied [44], the results of such experiments having relevance to the simple catalytic process whereby hydrogen is dissociatively chemisorbed onto a transition metal surface. An interesting feature of the reaction of iron clusters with molecular hydrogen or deuterium is the dependence of the rate of reaction on the size of the cluster, clusters of size $Fe_{15}$ to $Fe_{18}$ showing very low reactivity. Indeed abrupt changes in rate constant from one cluster to the next are seen, and the temperature dependence of these reactions is also complex. The more reactive clusters show decreased reactivity with increased temperature, while the least reactive clusters become more reactive as the temperature is raised. Clearly much more work is required in order to understand fully these interesting and complicated reactions.

The reactivity of small iron clusters with oxygen, hydrogen sulphide and methane has also been studied [45]. These experiments show that Fe clusters are very reactive towards $O_2$ and $H_2S$ yielding products of the type $Fe_xO_2$ and $Fe_xS$ (where $x = 2$–15), by molecular addition and dehydrogenation reactions, respectively. However, under identical conditions, these same iron clusters show no reaction with methane. It is found that the reactivity towards $O_2$ and $H_2S$ increases considerably for the smallest clusters, but levels off for clusters where $x > 6$.

## 4.11 METAL VAPOUR SYNTHESIS

To end this chapter, the use of metal atoms in *preparative* chemistry will be considered. Metal vapour synthesis may be achieved by co-condensing a metal vapour with a reactive ligand in a vessel held at liquid nitrogen temperature (96 K). After warm-up, the products of the metal–ligand reaction may be recovered. This is a very active area of chemistry, and there is much published work in this field. However, one recent example from the group of Malcolm Green at the University of Oxford will serve to illustrate the use of the technique to prepare new molecules and to investigate interesting chemical reactions. This example concerns the reactivity of rhenium atoms with various hydrocarbons [46].

It is found that co-condensation of rhenium atoms with certain alkylbenzenes gives dirhenium compounds containing bridging alkylidene ligands. For example, ethylbenzene yields a product containing the $Re(\mu\text{-}CHCH_2Ph)(\mu H_2)Re$ moiety in which the methyl group of ethylbenzene has added across the two Re atoms. On the other hand, Re atoms yield no isolatable product with benzene under these conditions. However, co-condensation of Re atoms with mixtures of benzene and saturated hydrocarbons, for example ethane or *neo*-pentane, gives products where activation of the alkane by C–H bond insertion has occurred. The structures of the products of reaction with ethane (**36**) and *neo*-pentane (**37**) are given below. Reactions of this sort, involving alkane activation, are of great interest, and the results of these experiments may be compared with the matrix-isolation experiments described in section 4.5.

Ozin and his co-workers have described an interesting variation on the metal vapour synthesis technique, which uses a helium refrigeration to condense, with a

(**36**)                                   (**37**)

metal vapour, reactive molecules which are non-condensable at liquid nitrogen temperature. He has adopted the title 'preparative matrix isolation' for this method by which he has succeeded in synthesizing the molecule ruthenium pentacarbonyl by co-condensation of ruthenium atoms with carbon monoxide [47]. This product is

$$Ru + 5CO \longrightarrow Ru(CO)_5$$

stable at room temperature and can be extracted into organic solvents. Fig. 4.13 illustrates the infrared spectrum obtained for the molecule $Ru(CO)_5$ in solution, and it shows also that, on standing, the molecule gradually trimerizes to the cluster $Ru_3(CO)_{12}$.

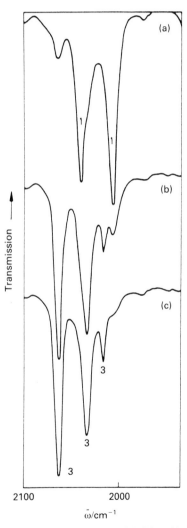

Fig. 4.13 — Product of the reaction of Ru atoms with CO at 30 K, extracted under argon into pentane at 200 K: (a) after 3 min; (b) after 196 min; (c) after 1478 min. Bands marked '1' belong to $Ru(CO)_5$ and bands marked '3' belong to $Ru_3(CO)_{12}$. Reproduced with permisssion from Godber et al., Inorg. Chem., **25**, 2909 © 1986 American Chemical Society.

**4.12 SUMMARY**

The sections in this chapter describe only a small part of the interesting chemistry displayed by metal atoms and small metal clusters. The results of these experiments complement, to some extent, those obtained from photochemical studies on metal carbonyl molecules described in Chapter 3. For example, unsaturated metal carbonyls may be prepared in low-temperature matrices either by photolysis of a stable metal carbonyl precursor, or by co-condensation of metal atoms with mixtures of carbon monoxide and an inert gas. It is interesting to compare the products formed by these 'breaking-down' and 'building-up' reactions. Similarly, other molecules such as metal oxides, metal dinitrogen complexes and metal alkene or alkyne complexes may be synthesized by photolytic or metal–atom reaction routes.

The structure and reactivity of small metal clusters are of considerable interest. Not only do such clusters provide us with information about the aggregation of metal atoms in the gas phase, during condensation of a metal, but they also provide possible models for heterogeneous catalysts. Information on these 'naked' metal clusters has come both from matrix isolation experiments and also from gas phase studies of supersonic molecular beams.

Lastly the preparative aspects of metal atom chemistry must be considered. Metal vapour synthesis is now an established and convenient route to a wide range of inorganic and organometallic molecules, and such experiments have provided much information on insertion reactions and the activation of saturated hydrocarbons. Ozin's preparative matrix isolation technique may well extend the scope of metal vapour synthesis, provided significant yields of product can be achieved in this way.

**REFERENCES**
[1]  R. L. De Kock, *Inorg. Chem.* (1971), **10**, 1205.
[2]  E. P. Kündig and G. A. Ozin, *J. Am. Chem. Soc.*, (1974), **96**, 3820.
[3]  R. N. Perutz and J. J. Turner, *Inorg. Chem.*, (1974), **14**, 262.
[4]  J. D. Black and P. S. Braterman, *J. Am. Chem. Soc.* (1975), **97**, 2908.
[5]  H. Huber, E. P. Kündig, G. A. Ozin and A. J. Poë, *J. Am. Chem. Soc.* (1975), **97**, 308.
[6]  L. A. Hanlan, H. Huber, E. P. Kundig, B. R. McGarvey and G. A. Ozin, *J. Am. Chem. Soc.* (1975), **97**, 7054.
[7]  A. Bos, *J. Chem. Soc., Chem. Commun.* (1972), 26.
[8]  A. J. Hinchcliffe, J. S. Ogden and D. D. Oswald, *J. Chem. Soc., Chem. Commun.* (1972), 338.
[9]  J. L. Slater, R. K. Sheline, K. C. Lin and W. Weltner, Jr., *J. Chem. Phys.*, (1971), **55**, 5129.
[10] J. S. Ogden, *J. Chem. Soc., Chem. Commun.* (1971), 978.
[11] D. McIntosh and G. A. Ozin, *J. Am. Chem. Soc.* (1976), **98**, 3167.
[12] D. McIntosh, M. Moskovits and G. A. Ozin, *Inorg. Chem.* (1976), **15**, 1669.
[13] L. Andrews and R. R. Smardzewski, *J. Chem. Phys.* (1973), **58**, 2258.
[14] L. Andrews, *J. Phys. Chem.* (1969), **73**, 3922.
[15] L. Andrews, *J. Am. Chem. Soc.* (1973), **95**, 4487.
[16] L. Andrews, J.-T. Hwang and C. Trindle, *J. Phys. Chem.* (1973), **77**, 1065.
[17] A. Bos and J. S. Ogden, *J. Phys. Chem.* (1973), **77**, 1513.

[18a] A. A. Malt'sev and L. V. Serebrennikov, *Vestn. Mosk. Univ. Khim.* (1975), **30**, 363.

[18b] J. H. Darling, M. B. Garton-Spencer and J. S. Ogden. *Symp. Faraday Soc.* (1974), **8**, 75.

[19] K. R. Thompson, W. C. Easley and L. B. Knight, *J. Phys. Chem.* (1973), **77**, 49; J. S. Shirk and A. M. Bass, *J. Chem. Phys.* (1970), **52**, 1894.

[20] H. Huber, W. Klotzbücher, G. A. Ozin and A. Vander Voet, *Canad. J. Chem.* (1973), **51**, 2722.

[21] D. H. W. Carstens, D. M. Gruen and J. F. Kozlowski, *High Temp. Sci.* (1972), **4**, 435.

[22] H. Huber, D. McIntosh and G. A. Ozin, *Inorg. Chem.* (1977), **16**, 975.

[23] G. A. Ozin, A. J. L. Hanlan and W. J. Power, *Inorg. Chem.* (1979), **18**, 2390.

[24] H. Huber, G. A. Ozin and W. J. Power, *J. Am. Chem. Soc.* (1976), **98**, 6508.

[25] P. H. Kasai and D. McLeod, Jr., *J. Am. Chem. Soc.* (1975), **95**, 5609.

[26] P. H. Kasai, D. McLeod, Jr. and T. Watanabe, *J. Am. Chem. Soc.* (1977), **99**, 3521.

[27] Z. H. Kafafi, R. H. Hauge and J. L. Margrave, *J. Am. Chem. Soc.* (1985), **107**, 7550; E. S. Kline, Z. H. Kafafi, R. H. Hauge and J. L. Margrave, *J. Am. Chem. Soc.* (1985), **107**, 7559.

[28a] W. E. Billups, M. M. Konarski, R. H. Hauge and J. L. Margrave, *J. Am. Chem. Soc.* (1980), **102**, 7393.

[28b] G. A. Ozin and J. G. McCaffrey, *J. Am. Chem. Soc.* (1982), **104**, 7351.

[29] Z. H. Kafafi, R. H. Hauge, L. Fredin, W. E. Billups and J. L. Margrave, *J. Chem. Soc., Chem. Commun.* (1983), 1230.

[30a] J. Chetwynd-Talbot, P. Grebenik and R. N. Perutz, *Inorg. Chem.* (1982), **21**, 3647.

[30b] D. M. Haddleton, *J. Organomet. Chem.* (1986), **311**, C21.

[31] J. W. Kauffman, R. H. Hauge and J. L. Margrave, *J. Phys. Chem.* (1985), **89**, 3541; J. W. Kauffman, R. H. Hauge and J. L. Margrave, *J. Phys. Chem.* (1985), **89**, 3547.

[32] M. Park, R. H. Hauge, Z. H. Kafafi and J. L. Margrave, *J. Chem. Soc., Chem. Commun.* (1985), 1570.

[33] S-C. Chang, Z. H. Kafafi, R. H. Hauge, W. E. Billups and J. L. Margrave, *J. Am. Chem. Soc.* (1985), **107**, 1447.

[34] L. Andrews and G. C. Pimentel, *J. Chem. Phys.* (1966), **44**, 2527; *J. Chem. Phys.* (1967), **47**, 3637.

[35] A. Snelson, *J. Phys. Chem.* (1970), **74**, 537.

[36] J. K. Burdett, *J. Mol. Spectrosc.* (1970), **36**, 365.

[37] D. P. DiLella, W. Limm, R. H. Lipson, M. Moskovits and K. V. Taylor, *J. Chem. Phys.* (1982), **77**, 5263.

[38] M. Moskovits, W. Limm and T. Mejean, *J. Chem. Phys.* (1985), **82**, 4875.

[39] H. Sontag, B. Eberle and R. Weber, *Chem. Phys.* (1983), **80**, 279; B. Eberle, H. Sontag and R. Weber, *Chem. Phys.* (1985), **92**, 417.

[40] D. A. Garland and D. M. Lindsay, *J. Chem. Phys.* (1983), **78**, 2813; *J. Chem. Phys.* (1984), **80**, 4761.

[41] J. A. Howard, K. F. Preston and B. Mile, *J. Am. Chem. Soc.* (1981), **103**, 6226; J. A. Howard, R. Sutcliffe and B. Mile, *J. Phys. Chem.* (1983), **87**, 2268.

[42] P. A. Montano and G. K. Shenoy, *Solid State Commun.* (1980), **35**, 53.

[43] M. Moskovits and J. E. Hulse, *J. Phys. Chem.* (1977), **81**, 2004.

[44] M. E. Geusic, M. D. Morse and R. E. Smalley, *J. Chem. Phys.* (1985), **82**, 590;
S. C. Richtsmeier, E. K. Parks, K. Liu, L. G. Pobo and S. J. Riley, *J. Chem. Phys.* (1985), **82**, 3659.

[45] R. L. Whetten, D. M. Cox, D. J. Trevor and A. Kaldor, *J. Phys. Chem.* (1985), **89**, 566.

[46] J. C. Green, M. L. H. Green, D. O'Hare, R. R. Watson and J. A. Bandy, *J. Chem. Soc., Dalton Trans.* (1987), 391.

[47] J. Godber, H. X. Huber and G. A. Ozin, *Inorg. Chem.* (1986), **25**, 2909.

# 5

# Divalent silicon chemistry

## 5.1 INTRODUCTION

Divalent silicon species, silylenes, are important intermediates in reactions of silicon compounds. Currently, there is considerable research activity in various aspects of silylene chemistry. In particular, the development of methods for the production of silylenes and structural and kinetic studies are of importance. Kinetic studies are of much interest in relation to analogous reactions of carbenes.

Originally it was thought that, while to some extent the chemistry of silylenes resembles that of carbenes, their reactivity is generally rather lower. However, recent real-time kinetic measurements have suggested that there is, in many cases, but little difference between the rates at which these two classes of species undergo analogous chemical reactions, although the kinetics of silylene reactions are often quite complicated. However some differences do exist. For example the silylene, dimethylsilylene (1) and its Si=C double-bonded isomer methylsilene (2), are almost isoenergetic and the isomerization (2→1) has been seen to occur either photochemically in matrices [1] or thermally in the gas phase. A similar isomerization of propene is obviously not a facile process.

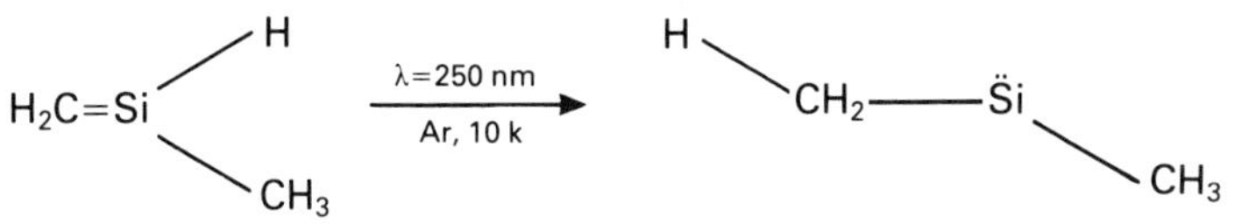

$$H_2C=Si\begin{matrix} H \\ \\ CH_3 \end{matrix} \quad \xrightarrow[\text{Ar, 10 k}]{\lambda=250 \text{ nm}} \quad \begin{matrix} H \\ \\ CH_2 \end{matrix}—\ddot{Si}\begin{matrix} \\ \\ CH_3 \end{matrix}$$

(2)                                                      (1)

Recently, considerable attention has been devoted to silylene chemistry, because of its relevance to the chemical vapour deposition (CVD) of polycrystalline silicon or amorphous hydrogenated silicon ($a$ − Si:H), the latter being potentially an important solar cell material. Most of this work has focused on the decomposition of silane ($SiH_4$), and the mechanisms proposed to account for silicon CVD are often based solely on gas chromatographic and mass spectroscopic analysis of the stable end products of $SiH_4$ decomposition. A fuller understanding of the reaction mechanism depends, however, on the detection of silylene intermediates and a knowledge of their chemical reactivity. This chapter describes a few of the methods by which silylenes have been produced and summarizes some of the techniques used for the detection of these short-lived species, including the measurement of kinetic and structural parameters.

## 5.2  PRODUCTION OF SILYLENES

Several methods are known by which silylenes can be generated. Often these involve decomposition of a stable silane precursor by pyrolysis, photolysis or the action of a microwave discharge. A different method, which is particularly suited to the matrix-isolation technique, is an insertion reaction of silicon atoms, e.g. into $H_2$ [2]:

$$Si + H_2 \xrightarrow[10\ K]{Ar\ matrix} H\ddot{S}iH$$

Pyrolysis methods have been applied successfully to the production of silylenes from both monosilanes and polysilanes. The gas phase pyrolysis of hexachlorodisilane ($SiCl_6$) in the presence of iodine yields dichlorodiiodosilane ($SiCl_2I_2$) as the major product. The kinetics of this reaction have been studied by Doncaster and Walsh, who have concluded that the following mechanism operates [3]:

$$Si_2Cl_6 \longrightarrow :SiCl_2 + SiCl_4$$

$$:SiCl_2 + I_2 \longrightarrow SiCl_2I_2$$

Thus the primary step of the reaction is the extrusion of dichlorosilylene, a process which has an activation energy, $E_a$, of $205.9 \pm 1.4\ kJ\ mol^{-1}$.

Gas phase pyrolysis of $SiH_4$ in a static system yields hydrogen and disilane as products. Two possible mechanisms are consistent with this observation, either a molecular elimination:

$$SiH_4 \xrightarrow{\Delta} :SiH_2 + H_2$$

$$:SiH_2 + SiH_4 \longrightarrow SiH_3SiH_3$$

or homolysis followed by radical chain reactions:

$$SiH_4 \xrightarrow{\Delta} \cdot SiH_3 + \cdot H$$

$$\cdot H + SiH_4 \longrightarrow H_2 + \cdot SiH_3$$

$$\cdot SiH_3 + SiH_4 \longrightarrow SiH_3SiH_3 + \cdot H$$

$$2\ \cdot SiH_3 \longrightarrow SiH_3SiH_3\ .$$

Over 20 years ago, Purnell and Walsh suggested that the former route was the most likely [4] and more recent experiments seem to have finally settled the question in favour of this molecular elimination pathway. Strausz and his co-workers carried out experiments in which silane was pyrolysed in the presence of 10% ethylene [5]. The presence of ethylene altered neither the product yields nor the rate of silane loss, an observation which effectively discounts the radical chain mechanism, since ethylene is known to be a scavenger of silyl radicals. Further experiments involving the decomposition of silane in a shock tube at temperatures between 1200 and 1300 K [6] suggest that the process has an activation energy of about 235 kJ mol$^{-1}$, a value which is much lower than the bond dissociation energy $D(SiH_3-H)$ of 393.3 kJ mol$^{-1}$, for $SiH_4$. Although this measured activation energy may not relate exactly to the primary step in silane decomposition, the results suggest that the product formed initially, on pyrolysis of silane, is likely to be silylene, $SiH_2$, rather than the silyl radical, $SiH_3$.

Silylenes can be generated by photolysis of polysilanes, a process which is related to the thermal decomposition of polysilanes mentioned above. It is usually described as the extrusion of the inner silylene unit:

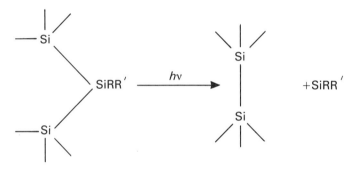

Photolysis of monosilanes has also received considerable attention. Lampe *et al.* have shown that irradiation of $SiH_4$ at 147 nm yields $SiH_2$ as the major primary product, while $SiH_3$ is formed by the minor primary reaction (b) [7]:

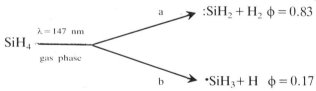

Among the final products are disilane and trisilane, and to account for this observation a mechanism involving the formation of disilane, in a vibrationally excited state, by both $SiH_2$ insertion and $SiH_3$ recombination was proposed:

$$:SiH_2 + SiH_4 \underset{\text{dissociation}}{\overset{\text{insertion}}{\rightleftharpoons}} SiH_3SiH_3^*$$

$$\cdot H + SiH_4 \longrightarrow H_2 + \cdot SiH_3$$

$$2\cdot SiH_3 \overset{\text{disproportionation}}{\longrightarrow} :SiH_2 + SiH_4$$

$$2\cdot SiH_3 \overset{\text{recombination}}{\longrightarrow} SiH_3SiH_3^*$$

$$SiH_3SiH_3^* + M \longrightarrow SiH_3SiH_3$$

$$SiH_3SiH_3^* \longrightarrow SiH_3\ddot{S}iH + H_2$$

$$SiH_3\ddot{S}iH + SiH_4 \xrightarrow{\text{insertion}} SiH_3SiH_2SiH_3^*$$

This reaction scheme, proposed by Lampe, illustrates the complexity of silane decomposition. Any process that converts silane to silyl radicals will also lead indirectly to silylene via disproportionation and recombination–dissociation reactions. In the case of the production of methyl-substituted silylenes from methyl silanes the situation is even more complicated. Formation of silylenes from dimethyl-silane, on photolysis at 147 nm, is an important reaction. Alexander and Strausz [8] have proposed the following mechanism for the photolysis of $(CH_3)_2SiD_2$. It is a complex mechanism, involving an array of reactive intermediates including silylenes, silyl radicals, alkyl radicals, carbenes and silenes.

$$(CH_3)_2SiD_2 \xrightarrow[\lambda = 147 \text{ nm}]{hv} \begin{cases} & \phi \\ CH_3SiD: + CH_3D & 0.15 \\ CH_3SiD: + \cdot CH_3 + \cdot D & 0.20 \\ SiD_2: + 2\cdot CH_3 & 0.08 \\ CH_2 = SiD_2 + CH_4 & 0.05 \\ CH_3SiD_2H + :CH_2 & 0.04 \\ (CH_3)_2Si: + D_2 & 0.07 \\ CHSiD_2 + \cdot CH_3 + H_2 & 0.04 \\ ChSiCh_3 + D_2 + H_2 & 0.07 \\ \cdot CH_2SiCH_3 + HD + \cdot D & 0.09 \\ \cdot CH_2SiD_2CH_3 + \cdot H & 0.08 \\ CH_2 = SiDCH_3 + HD & 0.19 \end{cases}$$

In the scheme, $\phi$ refers to the quantum yield for each process and it can be seen that about 50% of these primary reactions produce silylenes and about 25% C= Si-bonded silenes.

Finally photolysis of 1,1-dimethylsilacyclobutane, **3**, at wavelengths around 147 nm, yields 2-methyl-2-silapropene and ethylene as the major products, but also dimethylsilylene as a minor product [9].

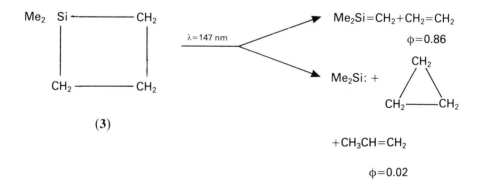

(**3**)

Silylene, SiH$_2$, has recently been detected in glow discharges in either silane or disilane [10]. This observation is of particular importance because such discharges are often used to initiate the deposition of amorphous hydrogenated silicon ($a-$Si:H) from silanes.

A totally different approach to the formation of unstable silicon compounds is the reaction of silicon atoms, with a number of different substrates. Co-condensation of silicon vapour with excess N$_2$ or CO at 4 K [11] leads to the formation of SiN$_2$ and SiCO; the products have been identified by electron spin resonance, ultraviolet and infrared spectroscopy. Annealing a CO matrix, containing silicon atoms, to 35 K leads to the formation of Si(CO)$_2$, an analogue of carbon suboxide, by the following reaction:

$$Si(CO) + CO \longrightarrow Si(CO)_2 \ .$$

Recent work has shown that silylene, SiH$_2$, itself can be produced in argon matrices at 10 K, by a similar reaction [2]. Silicon atoms will spontaneously insert into the H–H bond of molecular hydrogen, on co-condensation of silicon vapour, molecular hydrogen and an excess of argon at 10 K. A further insertion of SiH$_2$ into a second molecule of hydrogen yields silane, SiH$_4$.

$$Si + H_2 \ \xrightarrow[\text{10 K}]{\text{Ar matrix}} \ : SiH_2$$

$$SiH_2 + H_2 \longrightarrow SiH_4$$

Although this approach is well suited to the matrix-isolation technique, similar studies have been made, in other media, on reactions of recoiling silicon atoms. These recoiling atoms are generally produced by neutron irradiation of PH$_3$ or PF$_3$. Thus, in the presence of butadiene, $^{31}$Si atoms yield the spirocompound (**5**) as a minor product. This implicates the intermediacy of the cyclic silylene (**4**) in the reaction [12].

(**4**)                                          (**5**)

Similarly, neutron bombardment of PF$_3$ yield $^{31}$Si atoms which can abstract two fluorine atoms from their PF$_3$ precursor yield $^{31}$SiF$_2$ [13]. Because of the low concentrations of SiF$_2$ produced in this way, dimerization is not a major problem, so chemically stable products containing a single silylene unit predominate. This method has therefore proved to be very useful in studying reactions of monomeric SiF$_2$.

## 5.3  SILYLENE (SiH$_2$)

The first electronic spectrum of silylene was observed from the flash photolysis of phenylsilane. However, much improved spectra were soon obtained by Dubois,

Herzberg *et al.* in the 480–650 nm region, from silylene produced by a flash discharge through silane in excess hydrogen [14]. The spectrum mainly consists of a progression of seven bands, corresponding to the excitation of the bending vibration of the *upper* electronic state ($^1B_1$) of the $SiH_2$ radical. The spacing of these bands gives rise to a value of $\sim 860$ cm$^{-1}$ for the bending vibration ($v_2$) of the $^1B_1$ state of $SiH_2$. In addition to the main absorption, three weak bands were observed corresponding to hot bands from excitation of the bending vibration of the *lower* electronic state ($^1A_1$). From the spacing of these bands, a value of $\sim 1004$ cm$^{-1}$ was obtained for the bending vibration of the $^1A_1$ state. A rotational analysis was carried out on the fine structure of these bands. This yields H–$\hat{Si}$–H bond angles of 92° and 123° and $r_{Si–H}$ bond lengths of 1.516 and 1.487 Å for the $^1A_1$ and $^1B_1$ states of silylene, respectively. The geometry of silylene, as deduced from this study, is summarized in Table 5.1.

**Table 5.1** — Some measured parameters of gas phase $SiH_2$

| Electronic state | $r_{Si–H}$/Å | H–$\hat{Si}$–H | $v_2$/cm$^{-1}$ |
|:---:|:---:|:---:|:---:|
| $^1A_1$ | 1.516 | 92° | $\sim 1004$ |
| $^1B_1$ | 1.487 | 123° | $\sim 860$ |

Although it was clear that the observed spectrum arose from the $^1B_1 \leftarrow {}^1A_1$ transition, it was not certain whether the $^1A_1$ state was the ground state, or whether, as in methylene, $CH_2$, there was a lower-lying triplet state. However semi-empirical calculations, carried out around this time, had predicted that $SiH_2$ was a molecule with a $^1A_1$ ground state, which had a bond angle of around 95° [15]. More recent *ab initio* calculations have shown that the $^1A_1$ state is indeed the ground state, and that the lowest-lying triplet state, $^3B_1$, lies some 48–65 kJ mol$^{-1}$ above the $^1A_1$ state [16]. The geometries and vibrational frequencies of the $^1A_1$, $^3B_1$ and $^1B_1$ electronic states of $SiH_2$, calculated in this way, are listed in Table 5.2. As can be seen, these compare quite well with the experimentally obtained values listed earlier. However, an accurate measurement of the infrared spectrum of $SiH_2$ is of obvious importance, and a good way of achieving this is to generate $SiH_2$ in a low-temperature matrix.

**Table 5.2** — Some calculated parameters of $SiH_2$

| Electronic state | H–$\hat{Si}$–H | $v_1$/cm$^{-1}$ | $v_2$/cm$^{-1}$ | $v_3$/cm$^{-1}$ |
|:---:|:---:|:---:|:---:|:---:|
| $^1A_1$ | 97.46 | 2031 | 1132 | 2100 |
| $^3B_1$ | 123.53 | 2188 | 764 | 2218 |
| $^1B_1$ | 125.62 | 2019 | 751 | 2035 |

The first attempt to carry out such an experiment was by Milligan and Jacox, who subjected $SiH_4$ and $Si_2H_6$, isolated in argon matrices at 4 or 14 K, to vacuum ultraviolet photolysis [17]. They observed several products including, *inter alia*, SiH,

SiH$_2$ and SiH$_3$, which were identified by infrared and ultraviolet-visible spectroscopy. Unfortunately some of their assignments of bands were erroneous; it appears
that some of the infrared absorptions which they assigned to SiH$_2$ probably arose
rather from the radical SiH$_3$.

In a recent study, Fredin *et al.* have, however, succeeded in generating good
yields of SiH$_2$ in an argon matrix at 10 K by the spontaneous insertion reaction of a
silicon atom into the H–H bond of molecular hydrogen [2]. Isotopic studies were
carried out by substituting HD or D$_2$, for H$_2$, as the matrix dopant. As a result of
these studies the spectra illustrated in Fig. 5.1 were obtained and the values listed in
Table 5.3 for the $v_1$, $v_2$ and $v_3$ fundamentals of SiH$_2$, SiHD and SiD$_2$ were measured.

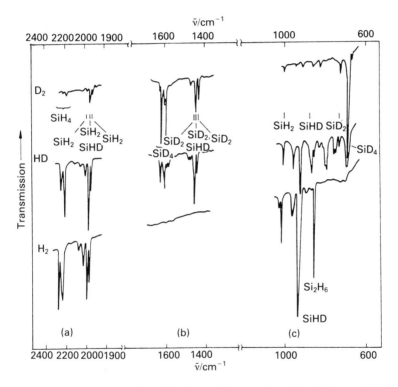

Fig. 5.1 — Infrared spectra of the products formed by co-condensation of Si atoms with H₂, HD
or D₂ and excess Ar. Reproduced with permission from Fredin *et al.*, *J. Chem. Phys.*, **82**, 3542
©1985 American Institute of Physics.

**Table 5.3** — Wavenumbers of vibrations of SiH$_2$, SiHD and SiD$_2$ in argon matrices at
10 K

| Vibrational mode | SiH$_2$ | SiHD | SiD$_2$ |
|---|---|---|---|
| $v_1$ | 1964.4 | 1973.3 | 1426.9 |
| $v_2$ | 994.8 | 854.3 | 719.8 |
| $v_3$ | 1973.3 | 1436.9 | 1439.1 |

The observed frequency of 994.8 cm$^{-1}$ for $v_2$ of SiH$_2$ compares quite closely with the value of 1004 cm$^{-1}$ calculated for the $^1A_1$ state of SiH$_2$ by Dubois [14]. Since only *ground* electronic states can be stabilized in matrices, this observation confirms the result of *ab initio* calculations [16], that the $^1A_1$ state is indeed the ground state of SiH$_2$.

Interestingly, silane, SiH$_4$, and disilane, Si$_2$H$_6$, were observed alongside SiH$_2$ as products of this reaction. Initially it was thought that the insertion reaction produced SiH$_2$ in an excited triplet state, and that this excited state molecule reacted with H$_2$ to form SiH$_4$. However, recent gas phase experiments have shown that there is no barrier to the reaction of ground-state SiH$_2$ and D$_2$ (see section 5.6.1) so it appears that the most likely reaction mechanism is as follows, in which the SiH$_2$ may undergo an insertion reaction with either H$_2$ or SiH$_4$ in the matrix to yield SiH$_4$ or Si$_2$H$_6$ respectively.

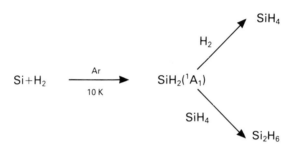

Two recent spectroscopic investigations of SiH$_2$ are of particular interest. These are the detection of SiH$_2$ in the gas phase, by laser-induced fluorescence [18], or by intracavity laser absorption spectroscopy [19].

Laser induced fluorescence has been observed from SiH$_2$ produced on photolysis of phenylsilane by an ArF excimer laser. After a short time delay ($\sim 20$ $\mu$s) the fragment, SiH$_2$, was excited by a tunable dye laser and fluorescence was detected by a photomultiplier. The time delay was introduced to avoid disturbance by the strong luminescence caused by the excimer laser protolysis. Fluorescence was seen from the transition from the excited $^1B_1$ state to the ground $^1A_1$ state of SiH$_2$. In these experiments the lifetime of the $^1B_1$ state, excited at 580.1 nm, was found to be $60 \pm 5$ ns, in 0.2 torr of helium. This lifetime is extremely short compared with that of $\sim 4000$ ns measured for the $^1B_1$ state of CH$_2$. Further information about SiH$_2$ is derived from studying the vibrational progression of the bands in its fluorescence spectrum; see Fig. 5.2. The vibrational spacing in the ground state is found to be $990 \pm 20$ cm$^{-1}$, a figure which is close to the value of 994.8 cm$^{-1}$ for the $v_2$ vibration of SiH$_2$, measured from its infrared spectrum [2].

The intracavity laser absorption spectrum (ILS) of SiH$_2$ has been observed by placing SiH$_4$ alongside argon in the cavity of a continuous-wave, jet-stream dye laser pumped by the 514.5 nm output of an argon ion laser. SiH$_2$ is generated *within* the laser cavity by the action of a microwave discharge, and the absorption spectra of intracavity molecular and atomic species are superimposed on the output of the laser. In this experiment the absorption spectrum of SiH$_2$ is superimposed upon those of H$_2$ and Ar. However, rotationally resolved spectra of the 020–000 and

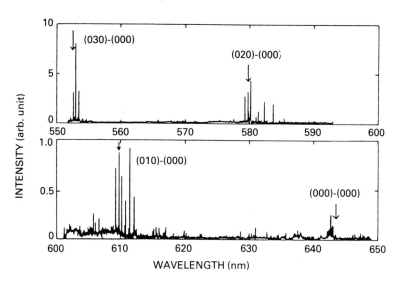

Fig. 5.2 — Laser-induced fluorescence spectrum of SiH$_2$ showing the band origin and vibrational progression. Reproduced with permission from Inoue and Suzuki, *Chem. Phys. Lett.*, **105**, © Elsevier 1984.

010–000 vibronic bands of the $^1B_1 \leftarrow {}^1A_1$ transition of SiH$_2$ can be identified. This work has suggested that there may be some mistaken assignments in the rotational analysis of the spectrum of SiH$_2$ carried out by Dubois [14]. One particular advantage of the ILS technique for studying SiH$_2$ is that it allows detection of SiH$_2$ at concentrations and under conditions (e.g. microwave discharge production) which are similar to the experimental conditions required for the chemical vapour deposition of Si and a-Si:H films. A diagram of the experimental arrangement required for the measurement of the ILS spectrum of SiH$_2$ is shown in Fig. 5.3.

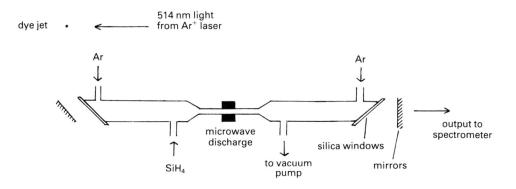

Fig. 5.3 — Diagram of the experimental arrangement required for measuring the intracavity laser spectrum (ILS) of SiH$_2$.

## 5.4  DIMETHYLSILYLENE (Me₂Si)

Dimethylsilylene, $Me_2Si$, had long been postulated as an organosilicon reaction intermediate before it was first observed directly by Drahnak, Michl and West in 1979 [20]. These workers found that irradiation of dodecamethylcyclohexasilane, **6** trapped in a rigid glass of 3-methylpentane or methylcyclohexane at 77 K, yielded decamethylcyclopentasilane, **7**, and dimethylsilylene by a photochemical extrusion reaction.

$$[(CH_3)_2Si]_6 \xrightarrow[\text{77k}]{\lambda=254\,nm} [(CH_3)_2Si]_5 + (CH_3)_2Si$$

**(6)**                                                        **(7)**

A similar reaction was observed in argon matrices at 10 K. The product, **7**, was characterized by comparison of its UV-visible and infrared spectra with those of an authentic sample of this stable compound. Dimethylsilylene was observed as a bright yellow species, with a broad visible absorption centred at 453 nm in 3-methylpentane glass, or 445 nm in an argon matrix and a characteristic infrared absorption at 1220 cm$^{-1}$. The visible absorption band is close, in energy, to those of other silylenes (e.g. $SiH_2$ at 480–650 nm). Furthermore, chemical trapping experiments provide convincing evidence that this yellow product is indeed $Me_2Si$. In these experiments a small quantity of the trapping agent is added as a dopant to the 3-methylpentane glass in which $Me_2Si$ is generated. Two of the chemical trapping agents employed were triethylsilane and 1-hexene. In the case of triethylsilane the insertion reaction

$$Me_2Si + Et_3SiH \longrightarrow Et_3Si\!-\!Si(H)Me_2$$

occurs. The disilane product can be recovered by melting the glass and separated by gas chromatography. It was characterized by proton NMR and mass spectrometry. In the case of 1-hexene being used as the trapping agent an addition reaction takes place:

yielding a silacyclopropane product which is treated with ethanol to give the expected ethanolysis product, hexadimethylethoxysilane, which was again separated by gas chromatography and characterized by mass spectroscopy.

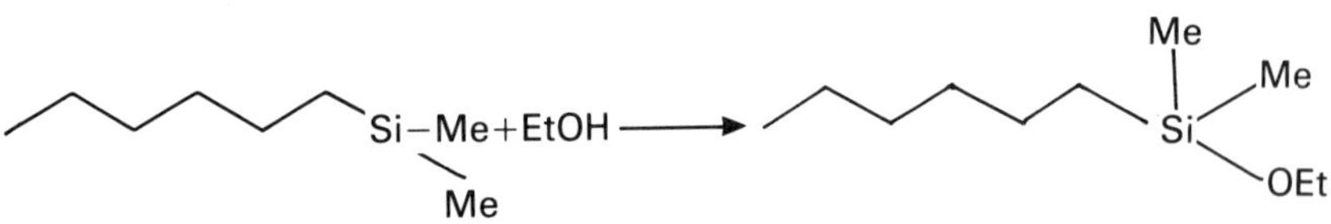

Yields in both cases were around 60%. Thus convincing evidence has been obtained for the existence of dimethylsilylene in solid matrices.

Although laser flash photolysis experiments on $(Me_2Si)_6$ in hydrocarbon solution gave rise to an absorption band close to 350 nm which was initially thought to belong to $Me_2Si$ [21], two recent theoretical studies, based either on substituent effects [22] or on *a priori* quantum mechanical methods [23], confirm that the 450 nm absorption seen in glasses and matrices is indeed due to ground-state $Me_2Si$. It seems possible that the 350 nm absorption seen in solution arises from an intermediate in the production of ground-state $Me_2Si$, such as a triplet state of $Me_2Si$, which would not be observed in low-temperature glasses or matrices. However, it is also possible that it arises from a weak complex formed between $Me_2Si$ and an impurity in solution.

A very interesting reaction is observed if matrix-isolated $Me_2Si$ is subjected to photolysis at wavelengths corresponding to its visible absorption at 450 nm. This is the isomerization to methyl silene $CH_3Si(H)=CH_2$, mentioned in the introduction to this chapter. The reaction can be reversed, either by ultraviolet photolysis at around 250 nm, or thermally. Since the thermal reaction will only occur at temperatures above 100 K it can only be observed in matrices which are involatile at this temperature, e.g. 3-methylpentane glass, and it appears that the process is mediated by the matrix material. The maximum temperature that an argon matrix can be warmed to is about 50 K, and at this temperature cyclodimerization of methylsilene predominates. These reactions are summarized in the scheme below:

Some elegant experiments involving polarized light for photolysis and polarized infrared spectroscopy have been carried out by the groups of Michl and West to probe this reaction [1]. Fig. 5.4 illustrates an infrared difference spectrum showing the conversion of matrix-isolated $Me_2Si$ to $Me(H)=CH_2$ by bleaching with visible light. Fig. 5.5 illustrates similar difference spectra, except that in this case the conversion has been effected by the Z-polarized 488 nm line of an $Ar^+$ ion laser, and the spectra have been recorded using radiation polarized along the Z-axis (parallel) and the Y-axis (perpendicular). After computer subtraction of absorptions due to other species present in the matrix the spectra have been plotted as the difference $E_Z - E_Y$ (i.e. the difference between extinction coefficients in the Z- and Y-axes). From these spectra it is possible to assign six IR transitions of $Me_2Si$ and 12 IR transitions of $CH_3Si(H)=CH_2$ as being in-plane or out-of-plane polarized. This has permitted detailed structural and vibrational assignments for the two molecules to be made.

Chemical trapping experiments have also yielded evidence in favour of this isomerization reaction [24]. Pyrolysis of 1-methylsilacyclobutane, **8**, in a tenfold

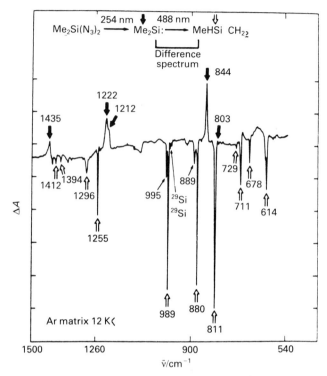

Fig. 5.4 — Infrared spectrum showing the photoconversion of $(CH_3)_2Si$ to $(CH_3)HSi = CH_2$. Reproduced with permission from Raabe *et al.*, *J. Am. Chem. Soc.*, **108**, 671 ©1986 Anmerican Chemical Society.

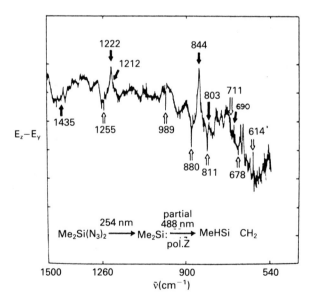

Fig. 5.5 — Infrared difference spectrum as in Fig. 5.4 except that the Z-polarized 488 nm line of an Ar⁺ laser has been used to effect the photoconversion. Reproduced with permission from Raabe *et al.*, *J. Am. Chem. Soc.*, **108**, 671 ©1986 American Chemical Society.

excess of butadiene at 600°C yields, *inter alia*, 1,1-dimethylsilacyclopent-3-ene, **9**, as a major product.

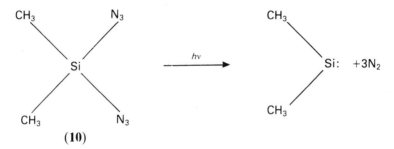

It is argued that pyrolysis of **8** initially yields methylsilene and ethylene

The methylsilene rapidly isomerizes to Me$_2$Si, at a temperature of 600°C, and the Me$_2$Si radical then adds to butadiene to yield the observed product **9**.

Recent efforts have been directed towards finding new precursors, from which high yields of dimethylsilylene can be produced. The polarization studies of Michl, West *et al.* were only possible because they obtained a high yield of Me$_2$Si by photolysing dimethyldiazidosilane, **10** [25].

Another precursor which gives a high yield of Me$_2$Si on photolysis is the trisilane, **11** [25].

$$PhMe_2Si-SiMe_2-SiPhMe_2$$

**(11)**

## 5.5   SILICON DIHALIDES (DIHALOSILYLENES, SiX$_2$)

The chemistry of silicon dihalides has received considerable attention recently because of its relevance to the plasma etching process. In this process, by which silicon is etched to form integrated circuit patterns, unmasked areas of a silicon substrate are subjected to attack by halogen atoms, which are produced in a plasma by the action of a glow discharge on a halogen-containing gas, e.g. $CF_4$. Material is thus removed in the form of volatile halides. While in the fluorine system, the final product is $SiF_4$, lower silicon fluorides, e.g. $SiF_2$, are observed on the substrate surface, and have been implicated as intermediates in the reaction mechanism.

Silicon difluoride was first made as an air-sensitive, rubbery polymeric material, $(SiF_2)_x$, by Schmeisser from magnesium and dibromodifluorosilane, $SiBr_2F_2$. It was later shown that silicon difluoride gas could be produced by passing $SiF_4$ at low pressure over silicon at 1150°C. This gas contains the monomeric $SiF_2$ radical which has a surprisingly long half-life, under these conditions, of around 2 min. Cooling the vapour to temperatures below $-80$°C resulted in the formation of the plastic polymer, $(SiF_2)_x$ [26].

A microwave study of $SiF_2$ in the gas phase has suggested a value of about 101° for the FŜiF bond angle, a bond length $r(Si-F)$ of 1.591 Å [27], and has predicted a value of 345 $cm^{-1}$ for the wavenumber of the $v_2$ bending vibration. Vibrational analysis of the ultraviolet absorption spectrum of gas phase $SiF_2$ confirms this value of 345 $cm^{-1}$ for $v_2$ of the ground singlet state of the molecule and also suggests a value of 252 $cm^{-1}$ for $v_2$ of the first excited singlet state [28]. Thus, like $SiH_2$, the bond angle of $SiF_2$ opens up considerably on the transition from the ground to the first excited singlet state. The geometry of the $^1A_1$ and $^1B_1$ states of silicon difluoride are summarized in Table 5.4.

**Table 5.4** — Some measured parameters of gas phase $SiF_2$

| Electronic state | $r_{Si-F}$/Å | F–Ŝi–F | $v_2$/$cm^{-1}$ |
|---|---|---|---|
| $^1A_1$ | 1.591 | 101° | 345 |
| $^1B_1$ | 1.49 | 124° | 252 |

The first attempt to record the ultraviolet spectrum of the $SiCl_2$ radical was made in 1938 by Asundi *et al.* [29]. They described a structured emission spectrum from flowing $SiCl_4$ in vapour discharges and assigned it to two band systems of the $SiCl_2$ molecule originating from excited states lying at 29952 and 28295 $cm^{-1}$ above the ground state. The higher lying of these states was later suggested to be the $^1B_1$ state, and the transition giving rise to one of the band systems to be $^1B_1 \rightarrow {}^1A_1$.

$SiCl_2$ has been produced in argon matrices by vacuum photolysis of $SiH_2Cl_2$ or $SiD_2Cl_2$ by Milligan and Jacox [30]. The infrared spectrum of this species was recorded and a value of 202 $cm^{-1}$ for the bending mode ($v_2$) of the ground state ($^1A_1$) was obtained. Recently the UV absorption spectrum of $SiCl_2$, produced by flash photolysis of $Si_2Cl_6$ in the gas phase, has been reported [31]. Analysis of the vibrational progression of this spectrum allows a value of 148 $cm^{-1}$ to be calculated for the bending mode of the $^1B_1$ excited state. The assignment of the spectrum to the $SiCl_2$ radical was confirmed by chemical scavenging experiments in which trans-2-butene was added to the reaction system. In these experiments gas-chromatographic separation, followed by mass spectrometric analysis, showed $C_4H_7^+$ and $SiCl_2H^+$ fragment ions to be present, an observation which is consistent with the following mechanism.

$$Si_2Cl_6 \xrightarrow{h\nu} :SiCl_2 \ + \ SiCl_4$$

:SiCl$_2$+ [alkene] $\longrightarrow$ [silirane with Si, Cl, Cl] $\longrightarrow$ [alkene] SiCl$_2$H + [alkene] SiCl$_2$H

The change in $v_2$ from 202 to 148 cm$^{-1}$ on moving from the ground, $^1A_1$, state to the excited, $^1B_1$, state is of a similar magnitude to the change from 345 to 252 cm$^{-1}$ observed for the corresponding states of SiF$_2$.

A study of the kinetics of the reaction between SiF$_2$ and F$_2$ has recently been made [32]. An understanding of the rate of this reaction is of importance because of its relevance to the plasma etching process, mentioned earlier. On the assumption that the products of the reaction between SiF$_2$ and F$_2$ would be SiF$_3$ and F, and modelling the conditions likely to be present during the plasma etching process, a half-time for SiF$_3$ formation of 0.02 to 0.2 s was estimated. These half-times are short enough to show that only SiF$_4$ would be present far from the substrate (as is observed), although they are long enough to suggest that SiF$_2$ participates in gas phase chemistry close to the silicon surface.

## 5.6 REACTIONS OF SILYLENES
The chemical reactions of silylenes may be considered under five separate headings. These are insertion, addition, abstraction, dimerization and rearrangement reactions. We will consider these five processes in turn.

### 5.6.1 Insertion reactions of silylenes
The most simple insertion reaction of a silylene is that between SiH$_2$, itself, and molecular hydrogen to form SiH$_4$. John and Purnell calculated Arrhenius parameters for this reaction [33] based on a combination of relative rate constants and an estimate of the heat of formation of SiH$_2$. They obtained an activation energy of ca. 26 kJ mol$^{-1}$. However, Jasinski has recently carried out experiments in which SiH$_2$ is generated by laser flash photolysis of phenylsilane, and the reaction between SiH$_2$ and D$_2$ monitored, in real time, by laser spectroscopy [34a]. The reaction scheme is as follows:

PhSiH$_3$ $\xrightarrow[\substack{\text{ArF excimer}\\\text{laser}}]{\text{193 nm}}$ SiH$_2$ $\underset{-D_2}{\overset{+D_2}{\rightleftarrows}}$ SiH$_2$D$_2$* 

SiH$_2$D$_2$* $\xrightarrow{-H_2}$ SiD$_2$

SiH$_2$D$_2$* $\xrightarrow{-HD}$ SiHD

SiH$_2$D$_2$* $\xrightarrow{M}$ SiH$_2$D$_2$

From these experiments a rate constant for the loss of $SiH_2$ by reaction with $D_2$ can be calculated. A value of ca. $2 \times 10^{-12}$ $cm^3$ $molecule^{-1}$ $s^{-1}$ is obtained for the reaction at room temperature, which is four orders of magnitude larger than the value of $2 \times 10^{-16}$ $cm^3$ $molecule^{-1}$ $s^{-1}$, which can be calculated from the Arrhenius parameters of John and Purnell. Thus it is suggested that the activation energy for the reaction between $SiH_2$ and $H_2$ is actually only about 4 kJ $mol^{-1}$, while other recent experiments have suggested that the reaction of $SiH_2$ with $D_2$ may proceed without an activation barrier [34b].

Insertion reactions of silylenes into several X–H bonds are known. A recent example is provided by the insertion reaction of $SiMe_2$ into the Si–H bonds of the silanes $SiH_4$, $MeSiH_3$, $Me_2SiH_2$ and $Me_3SiH$ [35]. The kinetics of this reaction have been followed by laser spectroscopy in real time, and an interesting observation is that the bimolecular reaction rate constant for the insertion of $SiMe_2$ into the Si–H bond appears to increase with increasing methyl group substitution on the silane. This is shown in Table 5.5, where $L$ is the reaction path degeneracy (i.e. $= 4$ for $SiH_4$; $= 3$ for $MeSiH_3$ etc.). These results show that the rate of reaction does not depend solely on the Si–H bond strength since experiments involving hydrogen atom abstraction from silanes by iodine atoms indicate that the strength of the Si–H bond is insensitive to methyl substitution of the Si atom [36].

**Table 5.5** — Relative rate constants for the reactions of $Me_2Si$ with various silanes

| Reaction | $10^{12}(k/L)/cm^3$ $molecule^{-1}$ $s^{-1}$ |
|---|---|
| $SiMe_2 + SiH_4$ | 0.05 |
| $SiMe_2 + MeSiH_3$ | 0.53 |
| $SiMe_2 + Me_2SiH_2$ | 2.75 |
| $SiMe_2 + Me_3SiH$ | 4.5 |

Silylenes will also insert into the O–H bonds of an alcohol, such as methanol. Theoretical calculations on these systems have indicated that these reactions are not a one-step process. Rather they involve initial formation of a donor-acceptor adduct between the silylene and the alcohol, followed by a 1,2-hydrogen shift. The calculations imply that there is a minimum in the potential energy surface of the reaction corresponding to the formation of the intermediate adduct.

Insertion reactions of silylenes are not limited to X–H bonds. Interesting reactions have been observed in which silylenes insert into C–O bonds, including those of strained cyclic ethers, such as oxetane. As in the case of O–H bond insertion reactions, it is assumed that these reactions proceed via a donor–acceptor complex, formed, in this case, by attack of the silicon atom on oxygen. It has been shown that photochemically generated dimethylsilylene reacts with oxetane at 0°C to give high yields of allyloxydimethysilane (**11**) and 2,2-dimethyl-1-oxa-2-silacyclopentane (**12**) [37a].

**11**                          **12**

At a temperature of $-98°C$, only product **11** is observed; this observation effectively discounts direct insertion as the mechanism for the formation of the cyclic product. It is proposed that the first step of the reaction is formation of an adduct which can undergo both intramolecular rearrangement to yield **11** and ring opening to a zwitterionic species followed by cyclization to yield **12**. However an alternative route could involve a diradical rather than a zwitterionic intermediate, such as that invoked in the insertion reaction of singlet $CH_2$ into a C–O bond of tetrahydrofuran. The two possible mechanisms are given below.

Both of these mechanisms are reasonable but more evidence is required to establish one of them firmly. However it is now becoming clear that silylenes will form adducts with oxygen containing molecules [37b]. For this reason, aliphatic ethers should not be used as solvents for silylene reactions, since these adducts may well have reactivities different from those of free silylenes.

### 5.6.2 Addition reactions of silylenes

Silylenes will add across carbon–carbon double bonds to generate siliranes–species with a three-membered C−Si−C ring. These siliranes are not, in themselves, stable products of the reaction; their presence is normally demonstrated by chemical trapping experiments in which methanol is used to produce a stable methoxysilane, which can then be characterized.

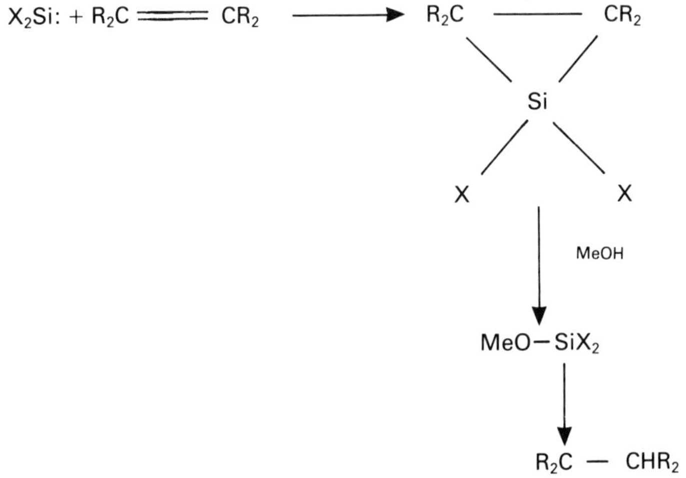

$$X_2Si: + R_2C =\!=\!= CR_2 \longrightarrow$$

R₂C ——— CR₂
\ /
Si
/ \
X      X

| MeOH

MeO−SiX₂

R₂C — CHR₂

Tortorelli and Jones carried out an experiment along these lines, in which they demonstrated that dimethylsilylene undergoes stereospecific addition to *cis*- and *trans*-2-butene. Dimethylsilylene was generated by photolysis of $(Me_2Si)_6$, and the products of its reaction with *cis*- and *trans*-2-butene were treated with methanol. In the $^1$H NMR spectrum, the methylene group of the resulting 2-butylmethoxydimethylsilane (**13**) appears as a diastereotopic pair of hydrogens Ha and Hb with chemical shifts $\delta = 1.18$ and 1.55 ppm [38].

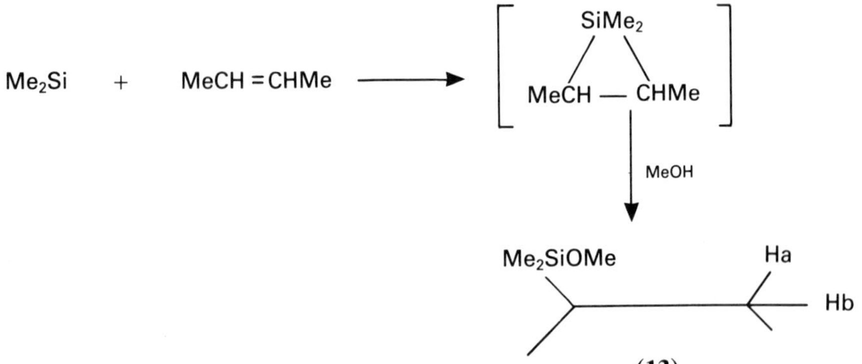

$$Me_2Si \quad + \quad MeCH = CHMe \longrightarrow \left[ \begin{array}{c} SiMe_2 \\ / \quad \backslash \\ MeCH — CHMe \end{array} \right]$$

| MeOH

Me₂SiOMe      Ha

Hb

(**13**)

However, if $CH_3OD$ is used in place of $CH_3OH$ only one of these hydrogens is seen in the $^1H$ NMR spectrum, while the other appears in the $^2H$ NMR spectrum. For *cis*-2-butene the deuterium atom resonates at $\delta = 1.18$ ppm and the hydrogen atom at 1.55 ppm, while for the *trans* isomer the deuterium atom resonates at $\delta = 1.55$ ppm and the hydrogen at $\delta = 1.18$ ppm. These observations demonstrate conclusively that only a single stereoisomer of the silirane is formed from the *cis* alkene and the other from the *trans* isomer. Importantly, the study also shows that methanolysis of siliranes is stereospecific.

This study has been extended to cover the reaction of dimethylsilylene and diphenylsilylene with various alkenes [39]. The experiments concerning the isomeric 2-butenes indicated, but did not prove, that the original mechanism involved concerted *cis* addition of the silylene. Although stereospecific *trans* addition would, in principle, provide the same result, such a process would be very unlikely. It is not, therefore, surprising that irradiation of $(Me_2Si)_6$ in methanol containing cyclopentane or cyclohexene yields the products **14** and **15**.

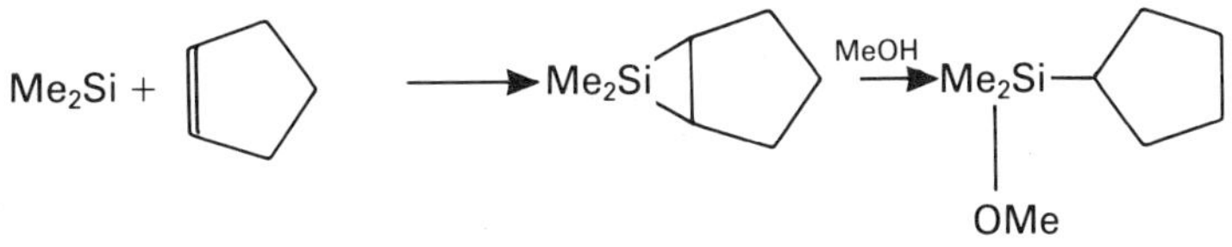

(**14**)

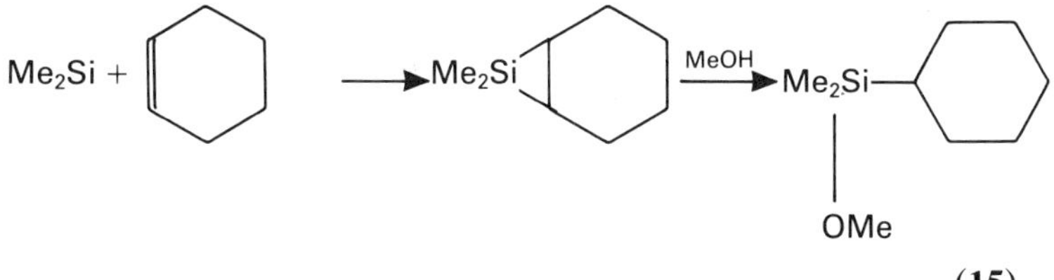

(**15**)

The structures of these products have been confirmed by comparison of NMR and IR spectra with samples of authentic compounds. Clearly, in these experiments *cis* addition must take place.

The addition of silylenes to conjugated dienes is rather more complex. In this case the reaction appears to proceed by a concerted 1,2-addition, followed by ring-opening of the vinylcyclopropane intermediate to diradicals which can then cyclize [40]. Thus the following products are obtained when $SiH_2$ reacts with *cis*, *trans*- or *trans*, *trans*-2,4-hexadiene.

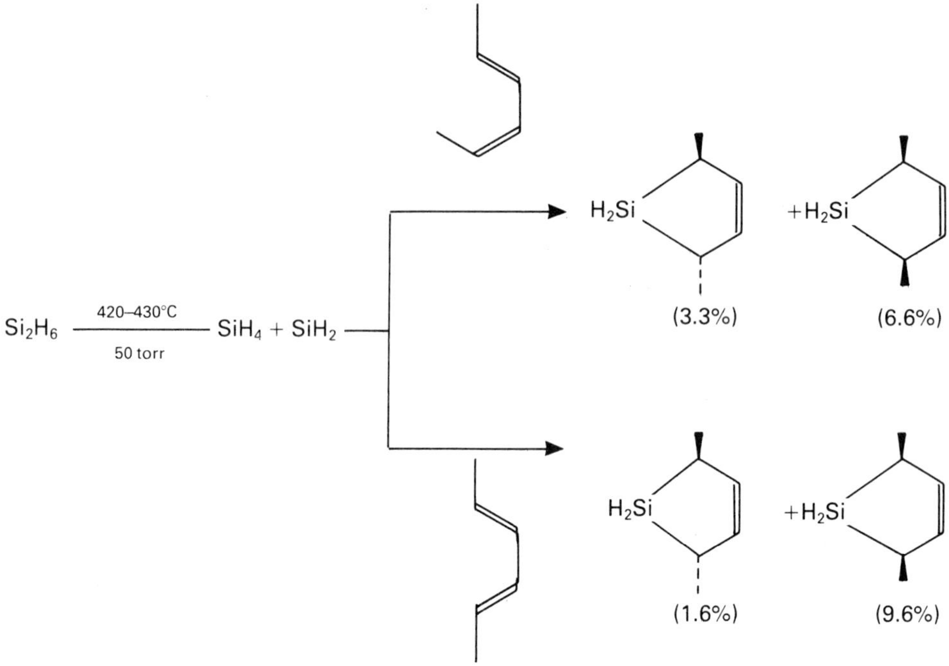

To account for this non-stereospecific addition the following mechanism has been proposed. This may, however, be oversimplified since there is a possibility of C–C as well as C–Si fission in the vinylcyclopropane intermediate, yielding other isomeric silacyclopentenes as products.

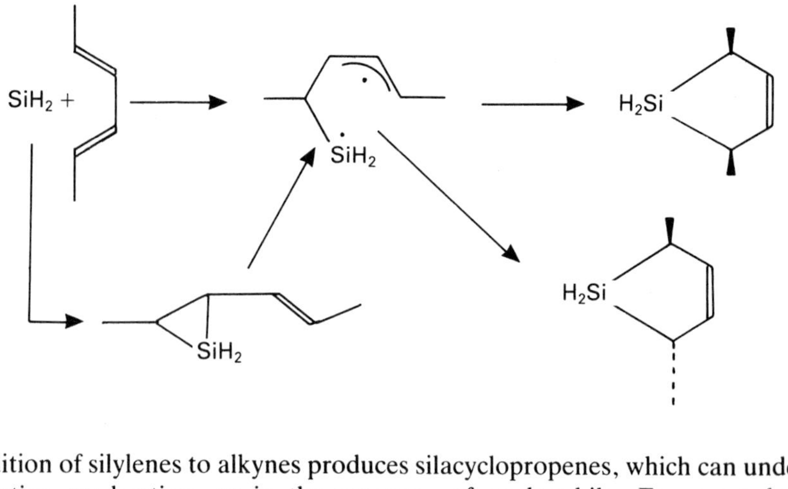

Addition of silylenes to alkynes produces silacyclopropenes, which can undergo dimerization on heating, or in the presence of nucleophile. For example the following process has been found which yields a dimer alongside a polymer [41].

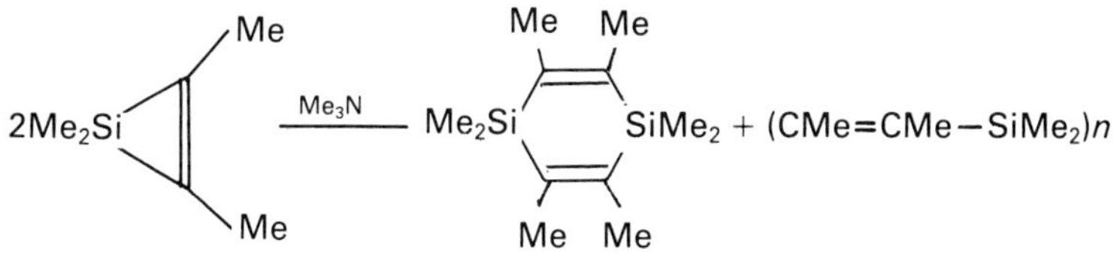

Alternatively the silacyclopropene may undergo ring-opening. Thus ethynylsilane is produced when thermally generated $SiH_2$ adds to acetylene itself.

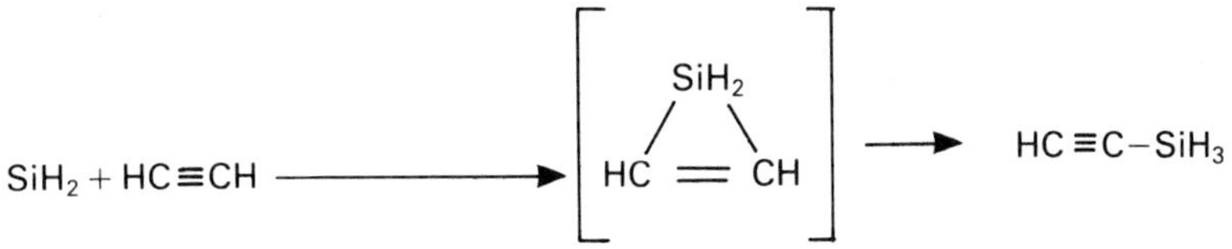

Recently, some addition reactions of silylenes with carbonyl compounds have been investigated. It is likely that such reactions proceed via oxysilirane intermediates which then undergo rearrangement. Two such examples are given below [42].

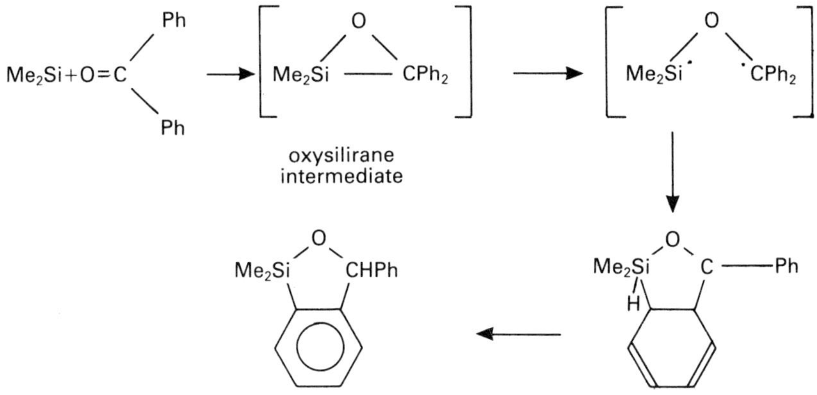

(16)

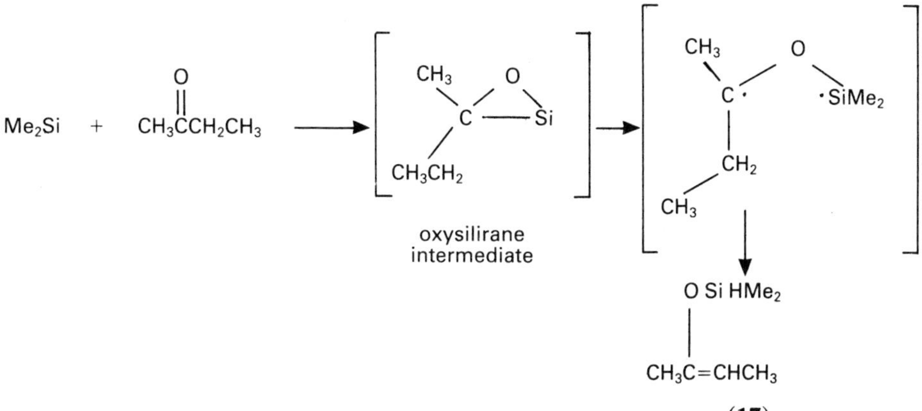

(17)

The cyclic sila-ether product (16) obtained from benzophenone and the enol-silaether product (17) from butanone are both products which are likely to be derived from oxysilirane intermediates. However, other possible mechanisms can be envisaged, for example one in which the initial step of the reaction is ylid formation, and there remain mechanistic questions about these reactions which still need to be answered.

### 5.6.3 Abstraction reactions of silylenes

An abstraction reaction of silylenes which has received some attention is the removal of an oxygen atom from dimethyl sulphoxide to form a silanone. In this, and other similar reactions, the silanone, itself, is a reactive intermediate, and its presence is normally inferred by chemical trapping experiments. For example, when dimethylsilylene is generated by photolysis of $(Me_2Si)_6$ in solutions containing dimethyl sulphoxide, dimethylsilanone is formed, and the presence of this product is indicated by its insertion into hexamethylcyclotrisiloxane [43].

$$(Me_2Si)_6 \xrightarrow{h\nu} (Me_2Si)_5 + Me_2Si$$

$$Me_2Si + O = SMe_2 \longrightarrow [Me_2Si=0] + SMe_2$$

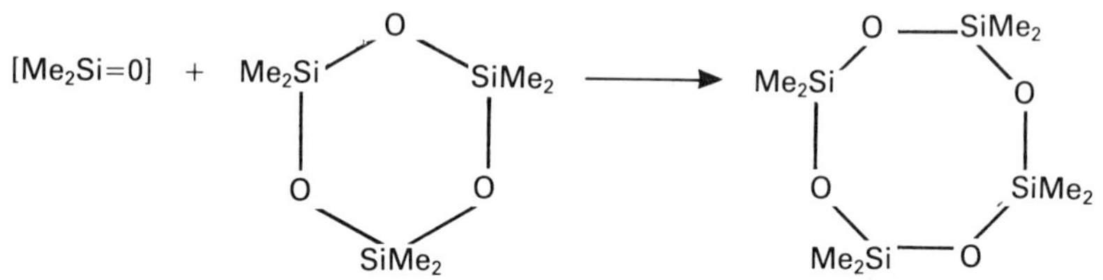

It appears that this deoxygenation reaction may well be a two-step process in which the first step is formation of an adduct between the silylene and dimethylsulphoxide. This intermediate may be compared with the adducts involved in the reactions between silylenes and ketones [37b].

$$R_2Si + O = SMe_2 \longrightarrow [R_2Si^- -O-S^+Me_2] \longrightarrow R_2Si = O + SMe_2$$

### 5.6.4 Dimerization reactions of silylenes

Dimerization reactions, which are almost unknown in carbene chemistry, are well established for organosilylenes. Sakurai et al. have obtained the expected anthracene Diels–Alder adducts from disilenes formed by dimerization of silylenes in solution. The reaction was carried out by cothermolysis of **18** and anthracene in a sealed tube at 350°C, and a 36% yield of **19** was obtained.

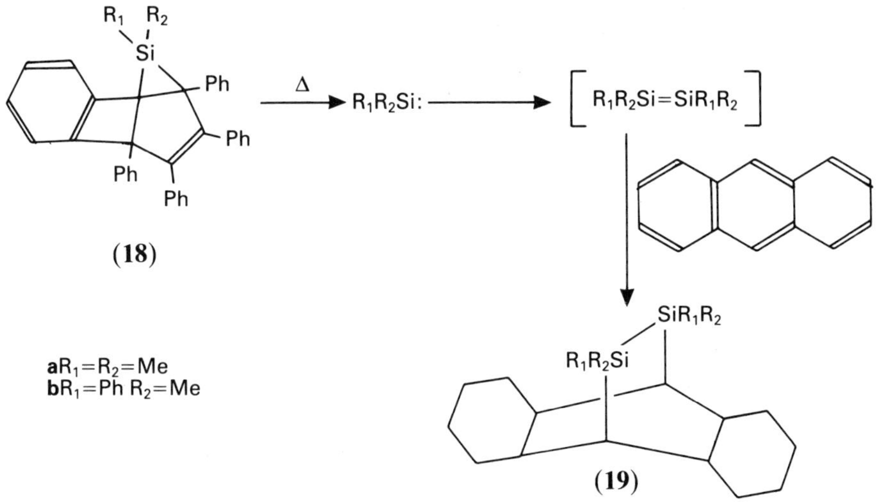

**(18)**

a$R_1 = R_2 = Me$
b$R_1 = Ph$ $R_2 = Me$

**(19)**

Interestingly, in the case of **18b** a 1:1 mixture of *trans* and *cis* isomers of the product **19b** was produced. Since *cis*- and *trans*-1,2-diphenyl-1,2-dimethyldisilene are configurationally stable, under these conditions, the result implies no stereochemical preference in the dimerization of phenylmethylsilylene.

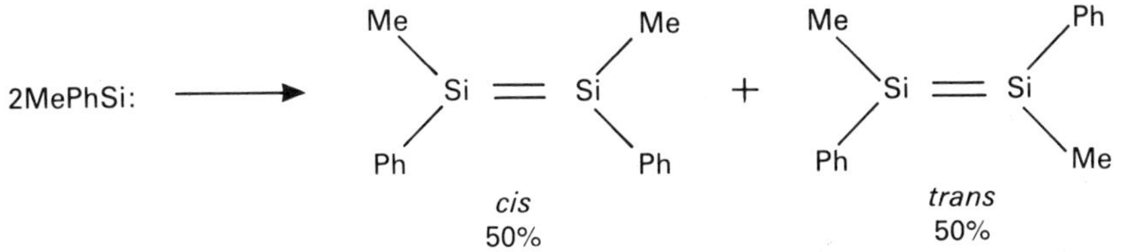

2MePhSi:

*cis*
50%

*trans*
50%

The same adduct **19b** is obtained if preformed 1,2-dimethyl-1,2-diphenyldisilene is generated in the presence of anthracene [44]. Thermolysis of *cis* or *trans* isomers of **20** in THF solutions containing anthracene yield, stereospecifically, the *cis* and *trans* isomers of **19b**.

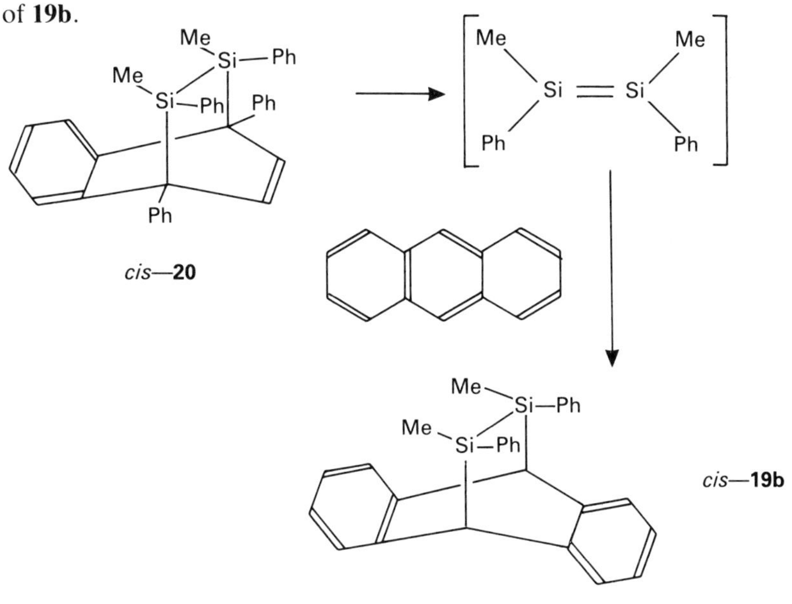

*cis*—**20**

*cis*—**19b**

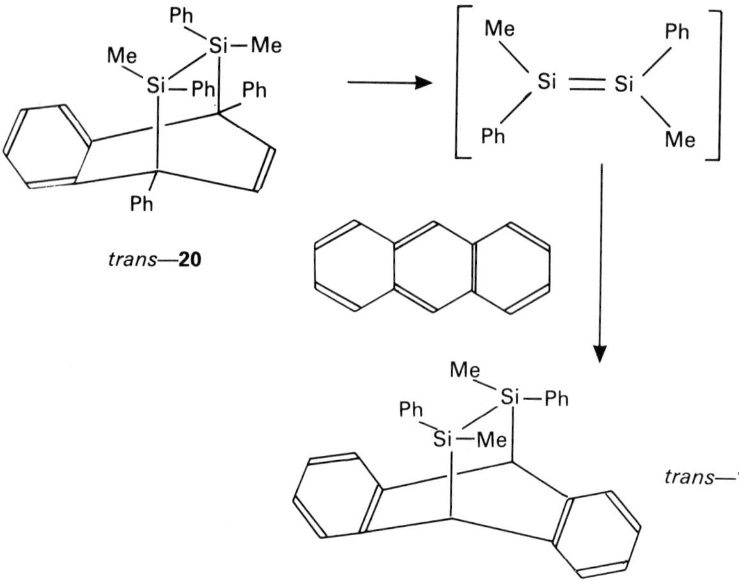

trans—**20**

trans—**19b**

The stereospecific formation of *cis* and *trans* isomers of the adduct **19b** demonstrates that under the experimental conditions the $\pi$ overlap between the silicon $3p$ orbitals is sufficient to retain the configuration of the disilene intermediate. In other words, the Si=Si bond is a true double bond. Indeed, the barrier to *cis–trans* isomerization has been estimated at 100–140 kJ mol$^{-1}$.

### 5.6.5   Isomerization reactions of silylenes

The silylene MeSiSiMe$_3$ will undergo dimerization, in solution, along the lines of the reactions described above. However, if this silylene is generated in the gas phase, by vacuum-flow pyrolysis, very different products are seen [45a]. Under these conditions the major products of the reaction are two isomeric disilacyclobutanes.

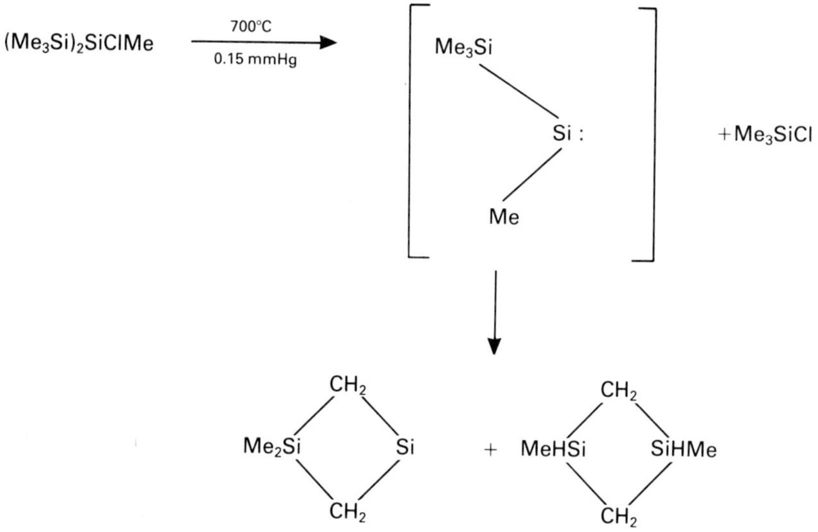

The formation of disilacyclobutane products indicates a silylene-to-silylene rearrangement of the silylene intermediate, analogous to the carbene-to-carbene rearrangements known, for example, for phenylcarbenes and vinylmethylenes. A possible explanation [45b] for the observation of two isomeric products is that $Me_3SiSiMe$ can suffer one of two fates: either (i) an intramolecular C−H insertion, yielding a highly strained disilacyclopropane, which then undergoes ring opening, or (ii) a silylene–silene isomerization, followed by migration of an $Me_3Si$ group. These reactions yield isomeric silylenes; further intramolecular C−H insertions will then produce the isomeric disilacyclobutane products.. The reaction mechanism is summarized below. It has the advantage over some earlier interpretations of excluding some rather unlikely methyl group migrations.

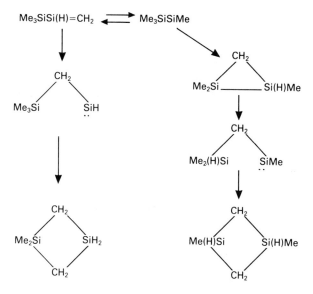

These products of the gas phase isomerization of $MeSiSiMe_3$ are the same as those which had previously been found following the extrusion of $Me_2Si=SiMe_2$ from **21**, in the absence of a trapping agent [46].

Thus Barton made the suggestion that tetramethyldisilene can rearrange to methyl-trimethylsilylene, a reverse of the usual carbene rearrangement.

$$Me_2Si = SiMe_2 \xrightarrow{\wedge} Me\ddot{S}i - SiMe_3$$

Similar isomerizations have now been observed for silenes, containing C=Si bonds, which form silylenes in the gas phase, at high temperatures:

That such reactions are only seen at high temperatures indicates a substantial activation barrier [47]. The isomerization of the Si=C bonded methylsilene to dimethylsilylene in an argon matrix at 10 K was mentioned in the introduction to this chapter; under these conditions, however, the reaction is promoted by photolysis at wavelengths close to 250 nm [1].

## 5.7  SUMMARY

This chapter provides an outline of the techniques employed to study silylenes, and describes some of the results of these studies. Originally much of the information which was obtained for the reactions of these species came from indirect kinetic information, derived, for example, from relative rate constants and estimates of thermochemical parameters, including heats of formation. This has led to some possible misconceptions regarding silylene chemistry, for example, concerning the relative reactivity of silylenes and the corresponding carbenes.

However, in recent years two types of measurement have become increasingly important. The first is direct spectroscopic study of silene intermediates trapped in low-temperature matrices. Secondly, kinetic parameters have been derived from real-time measurements, e.g. measurements of the decay of silylene intermediates, in laser flash-photolysis experiments, by laser spectroscopy.

Obviously difficulties will still arise. For example, in matrix-isolation experiments the possible influence of the matrix material must not be underestimated. This is exemplified by the thermal isomerization of methylsilene (Me(H)Si=CH$_2$) to dimethylsilylene (Me$_2$Si) which appears to proceed in hydrocarbon glasses at 100 K. Gas phase studies have shown that this process has a substantial activation barrier, so it is concluded that the process in the glass matrix must be promoted by the matrix material itself, or some impurity.

However, if this example shows a possible difficulty in interupting matrix-isolation results, It also shows how the problem can be solved by a combination of low-temperature matrix and gas phase studies. Together these techniques provide a powerful means of studying structure and reactivity, and they will be increasingly used together to solve problems in this rapidly expanding area of chemistry.

**REFERENCES**

[1] G. Raabe, H. Vancik, R. West and J. Michl, *J. Am. Chem. Soc.* (1986), **108**, 671.

[2] L. Fredin, R. H. Hauge, Z. H. Kafafi and J. L. Margrave, *J. Chem. Phys.* (1985), **82**, 3542.

[3] A. M. Doncaster and R. Walsh, *J. Chem. Soc., Faraday Trans. I* (1980), 272.

[4] J. H. Purnell and R. Walsh, *Proc. Roy. Soc. Lond., Ser. A* (1966), **293**, 543.

[5] P. Neudorfl, A. Jodhan and O. P. Strausz, *J. Phys. Chem.* (1980), **84**, 338.

[6] C. G. Newman, M. A. Ring and H. E. O'Neal, *J. Am. Chem. Soc.* (1978), **100**, 5945.

[7] G. G. A. Perkins, E. R. Austin and F. W. Lampe, *J. Am. Chem. Soc.* (1979), **101**, 1109.

[8] A. G. Alexander and O. P. Strausz, *J. Phys. Chem.* (1976), **80**, 2531.

[9] S. Tokach, P. Boudjouk and R. D. Koob, *J. Phys. Chem.* (1978), **82**, 1203.

[10] J. M. Jasinski, E. A. Whittaker, G. C. Bjorklund, R. W. Dreyfus, R. D. Estes and R. E. Walkup, *Appl. Phys. Lett.* (1984), **44**, 1155.

[11] R. R. Lembke, R. F. Ferrante and W. Weltner, Jr., *J. Am. Chem. Soc.* (1977), **99**, 416.

[12] R. J. Hwand and P. P. Gaspar, *J. Am. Chem Soc.* (1978), **110**, 6626.

[13] R. A. Ferrieri, E. E. Siefert, M. J. Griffin, O. F. Zeck and Y.-N. Tang, *J. Chem. Soc., Chem. Commoun.* (1977), **6**; E. E. Siefert, R. A. Ferrieri, O. F. Zeck and Y.-N. Tang, *Inorg. Chem.* (1978), **17**, 2802.

[14] I. Dubois, G. Herzberg and R. D. Verma, *J. Chem. Phys.* (1967), **47**, 4262; I. Dubois, *Can. J. Phys.* (1968), **46**, 2485.

[15] P. C. Jordan, *J. Chem. Phys.* (1966), **44**, 3400.

[16] B. Wirsam, *Chem. Phys. Lett.* (1962), **14**, 214.

[17] D. E. Milligan and M. E. Jacox, *J. Chem. Phys.* (1970), **52**, 2594.

[18] G. Inoue and M. Suzuki, *Chem. Phys. Lett.* (1984), **105**, 641.

[19] J. J. O'Brien and G. H. Atkinson, *Chem. Phys. Lett.* (1986), **130**, 321.

[20] T. J. Drahnak, J. Michl and R. West, *J. Am. Chem. Soc.* (1979), **101**, 5427.

[21] A. S. Nazran, J. A. Hawari, D. Griller, I. S. Alnaimi and W. P. Weber, *J. Am. Chem. Soc.* (1984), **106**, 7267.

[22] Y. Apeloig and M. Karni, *J. Chem. Soc., Chem. Commun.* (1985), 1048.

[23] R. S. Grev and H. F. Schaeffer, III, *J. Am. Chem. Soc.* (1986), **108**, 5804.

[24] R. J. Conlin and Y.-W. Kwak, *Organometallics* (1984), **3**, 918.

[25] H. Vancik, G. Raabe, M. J. Michalczyk, R. West and J. Michl, *J. Am. Chem. Soc.* (1985), **107**, 4097.

[26] P. L. Timms, R. A. Kent, T. C. Ehlert and J. L. Margrave, *J. Am. Chem. Soc.* (1965), **87** 2824.

[27] V. M. Rao, R. F. Curl, P. L. Timms and J. L. Margrave, *J. Chem. Phys.* (1965), **43**, 2557.

[28] V. M. Khanna, G. Besenbruch and J. L. Margrave, *J. Chem. Phys.* (1967), **46**, 2310.

[29] R. K. Asundi, M. Karim and R. Samuel, *Proc. Phys. Soc.* (*London*) (1938), **50**, 581.

[30] D. E. Milligan and M. E. Jacox, *J. Chem. Phys* (1968), **49**, 1938.

[31] B. P. Ruzsicska, A. Jodhan, I. Safarik, O. P. Strausz and T. N. Bell, *Chem. Phys. Lett.* (1985), **113**, 67.

[32] A. C. Stanton, A. Freedman, J. Wormhoudt and P. P. Gaspar, *Chem. Phys. Lett.* (1985), **122**, 190.

[33] P. John and J. H. Purnell, *J. Chem. Soc., Faraday Trans. I* (1973), **69**, 1445.

[34a] J. M. Jasinski, *J. Phys. Chem.* (1986), **90**, 555.

[34b] J. E. Baggott, H. M. Frey, K. D. King, P. D. Lightfoot and R. Walsh, unpublished results.

[35] J. E. Baggott, M. A. Blitz, H. M. Frey, P. D. Lightfoot, and R. Walsh, *Chem. Phys. Lett.* (1987), **135**, 39.

[36] R. Walsh, *Acc. Chem. Res.* (1981), **14**, 246.

[37a] T. Y. Y. Gu and W. P. Weber, *J. Am. Chem. Soc.* (1980), **102**, 1641.

[37b] W. Ando, K. Hagiwara and A. Sekiguchi, *Organometallics* (1987), **6**, 2270.

[38] V. J. Tortorelli and M. Jones, *J. Am. Chem. Soc.* (1980), **102**, 1425.

[39] V. J. Tortorelli, M. Jones, Jr., S. H. Wu and Z. H. Li, *Organometallics* (1983), **2**, 759.

[40] D. Lei, R. J. Hwang and P. P. Gaspar, *J. Organomet. Chem.* (1984), **271**, 1.

[41] C. H. Haas and M. A. Ring, *Inorg. Chem.* (1975), **14**, 2253.

[42] W. Ando and M. Ikeno, *Chem. Lett.* (1978), 609.

[43] Y. Nakadaira, T. Kobayashi, T. Otsuka and H. Sakurai, *J. Am. Chem. Soc.* (1979), **101**, 486.

[44] H. Sakurai, Y. Nakadaira and T. Kobayashi, *J. Am. Chem. Soc.* (1979), **101**, 487.

[45a] W. D. Wulff, W. F. Goure and T. J. Barton, *J. Am. Chem. Soc.* (1978), **110**, 6236.

[45b] I. M. T. Davidson, K. J. Hughes and R. J. Scampton, *J. Organomet. Chem.* (1984), **272**, 11.

[46] R. T. Conlin and P. P. Gaspar, *J. Am. Chem. Soc.* (1976), **98**, 868.

[47] R. Walsh, *J. Chem. Soc., Chem. Commun.* (1982), 1415.

# 6

# Aspects of organic photochemistry

## 6.1 INTRODUCTION

Most of this book is concerned with *inorganic* short-lived molecules. However, the traditional boundary between inorganic and organic chemistry is often not very distinct, and it is pertinent here to consider briefly the chemistry of some unstable organic molecules. In particular the following chapter describes some of the chemistry of divalent carbon compounds — the carbenes. There is an obvious comparison to be drawn between these species and the divalent silicon compounds — silylenes — discussed in Chapter 5. However, carbenes are also related, via the isolobal relationship, to unsaturated metal carbonyl fragments. Thus an analogy may also be drawn between carbene chemistry and the reactions of metal carbonyls, which were discussed in Chapter 3. Carbenes and metal carbonyls appear to be far-removed from one another, but some remarkable similarities in their chemistry have been found.

## 6.2 CARBENES (METHYLENES)

Carbenes are divalent carbon compounds, which may be produced by fragmentation reactions. Thus methylene, $CH_2$, is formed as the primary reaction product when diazomethane, $CH_2N_2$, is subjected to photolysis or pyrolysis in the gas phase. Methylene is, however, extremely reactive. It readily reacts with more diazomethane to yield ethylene, but will also react with the ethylene product, by insertion, to form propylene. Thus the principal stable reaction products are ethylene (63%), propylene (7%) and but-1-ene (9%).

$$CH_2N_2 \xrightarrow[\Delta]{h\nu \text{ or}} CH_2 + N_2$$

$$CH_2 + CH_2N_2 \rightarrow [CH_2{=}CH_2]^* + N_2 \rightarrow CH_2{=}CH_2$$

$$H_2C{=}CH_2 + CH_2 \xrightarrow{\text{insertion}} H_2C{=} C(H)CH_3, \text{ etc.}$$

Methylene is an electron-deficient species in which the carbon atom is associated with only six valence electrons. Two electronic structures may be proposed — singlet or triplet. The singlet structure (1) in which both electrons are paired in the same orbital, uses $sp^2$ hybrid orbitals for bonding, so the H−Ĉ−H bond angle should approximate to 120°. The triplet structure (2), with two unpaired electrons (each in a carbon $2p$ orbital), uses $sp$ hybrid orbitals for bonding. The geometry should, therefore, be much closer to linear.

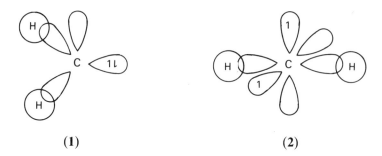

(1)                                        (2)

Gas phase photolysis of diazomethane yields both singlet and triplet methylene. However, in the presence of an inert gas, collisions convert the singlet to the triplet species. So unlike silylene, $SiH_2$, it appears that methylene has a triplet ground state. The energy difference between the triplet ground state and the lowest-lying singlet state of methylene has been calculated to be about 10 kJ mol$^{-1}$. Likewise, simple alkyl derivatives of methylene appear also to have triplet ground states.

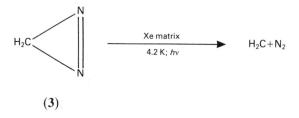

(3)

That methylene has a triplet ground state has been confirmed by the results of matrix-isolation experiments. Methylene has been produced by photolysis of diazirene (3) in a xenon matrix at 4.2 K [1], and the product detected by ESR spectroscopy. The observation of the ESR spectrum, illustrated in Fig. 6.1, demonstrates the presence of unpaired electrons in *ground-state* $CH_2$, since only ground electronic states are stabilized in low-temperature matrices. Measurement of the ESR spectrum of the isotopomer $^{13}CD_2$ has allowed the hyperfine interaction of the unpaired electron spin with the $^{13}C$ nucleus to be calculated. This gives a measure of the $s$ orbital character of the unpaired electron spin, and can be used to estimate carbon bond angles. Hence the H−Ĉ−H bond angle for triplet ground state $CD_2$ is calculated to be 137.7°, a value in line with theoretical estimates of 132–138° [2]. So the molecule does not have the linear structure predicted by the simple orbital arguments given earlier.

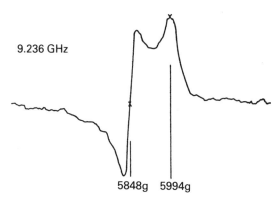

9.236 GHz

5848g   5994g

Fig. 6.1 — ESR spectrum of $CH_2$ produced by photolysis of diazirine in a xenon matrix. Reproduced with permission from Bernheim *et al.*, *J. Chem. Phys.*, **53**, 1280 © 1970 American Institute of Physics.

By contrast, dibromomethylene and other halogenocarbenes appear to have singlet ground states. Dibromomethylene has been prepared by co-condensation of $CBr_4$ and lithium atoms with excess argon at 15 K [3]. Values of 640.5 and 595.0 cm$^{-1}$ are recorded for $v_{asym}$ and $v_{sym}$ of $CBr_2$. The observation that both

$$CBr_4 + Li \xrightarrow[15k]{Ar} CBr_3 + LiBr$$

$$\downarrow Li$$

$$B_2Br_4 \xleftarrow{anneal} CBr_2 + LiBr$$

stretching modes are infrared active demonstrates that $CBr_2$ is a non-linear molecule of $C_{2v}$ symmetry, and a $Br-\hat{C}-Br$ bond angle of ~ 100° has been calculated for this molecule which has a singlet electronic ground state. Confirmation that these infrared bands do indeed arise from the molecule $CBr_2$ is obtained from annealing experiments. Under these conditions the absorptions of $CBr_2$ decay, while those of the known stable molecule $C_2Br_4$ appear and grow, presumably as a result of a dimerization reaction.

Many reactions of carbenes resemble those of silylenes, though the products may depend on whether the carbene is in a singlet or a triplet electronic state. Both singlet and triplet methylene will insert into C−H bonds. Singlet methylene reacts with hydrocarbons indiscriminately, inserting into all possible C−H bonds to give a near random distribution of products. On the other hand, triplet methylene is less reactive, and more selective in its action. It attacks tertiary C−H bonds about four times faster than primary ones. Addition reactions of both singlet and triplet carbenes to C=C double bonds are known; a valuable diagnostic technique to test for the presence of carbenes in a reaction mixture is trapping by means of an alkene. Singlet carbenes add by a concerted stereospecific cyclo-addition, groups *cis* to one another in the alkene appear *cis* also in the cyclopropane product — compare with the addition reactions of triplet silylenes described in section 5.6.2. An example is provided by the addition of methylene itself to *cis*-but-2-ene. Triplet methylene, on the other hand, adds to *cis*-but-2-ene with loss of stereochemical integrity.

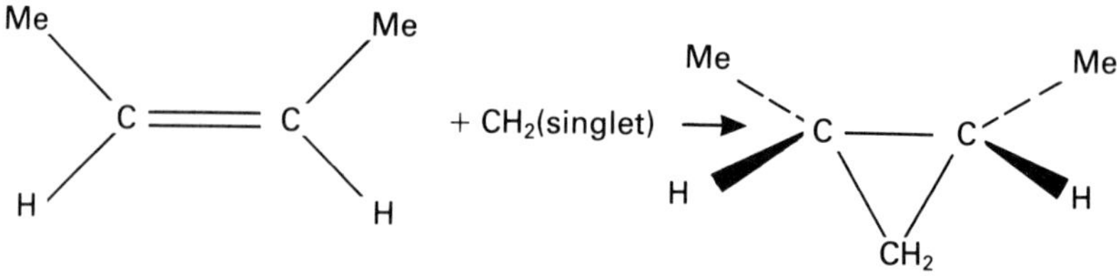

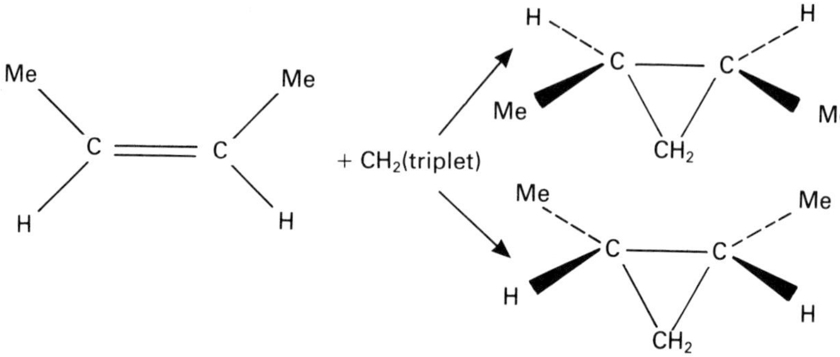

Rearrangement reactions of carbenes, including migration of hydrogen atoms or small organic groups, are well known, but dimerization reactions appear to be very rare. One example of a likely genuine dimerization, which was mentioned earlier, is the formation of $C_2Br_4$ from matrix-isolated $CBr_2$ on annealing [3].

Carbenes are related to unsaturated metal carbonyl fragments by the 'isolobal' relationship [4]. Two molecular fragments are said to be isolobal if the number, symmetry properties, shapes, and approximate energies of their frontier orbitals are the same. In this way methylene is related to the $d^8$ fragment $Fe(CO)_4$, which also has a triplet ground state (see section 3.6). So we might expect carbenes to show some chemical reactions, such as addition to CO or $O_2$, which are also known for $Fe(CO)_4$ [5].

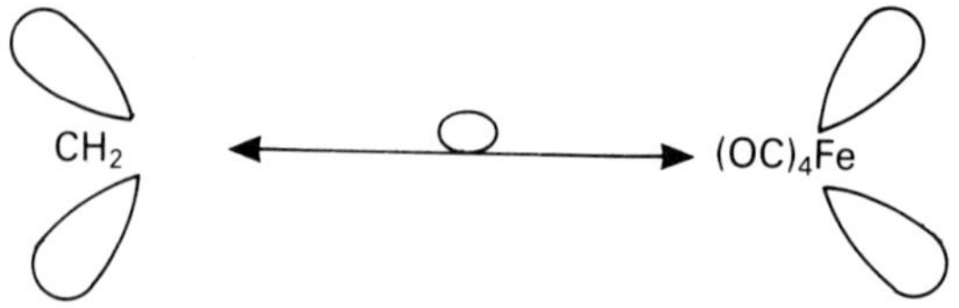

Matrix-isolated and gas phase carbenes have been shown to undergo simple addition reactions with CO to yield organic carbonyl compounds [6]. Of particular interest, however, are the reactions between carbenes and $O_2$, partly because the reaction products — carbonyl oxides — have been postulated as intermediates in the Criegee mechanism of alkene ozonolysis. However, their high reactivity has made their physical characterization difficult.

In 1983, Bell and Dunkin [7] reported that the carbonyl oxide cyclopentadienone O-oxide (**6**) is produced when diazocyclopentadiene (**4**) is subjected to photolysis in

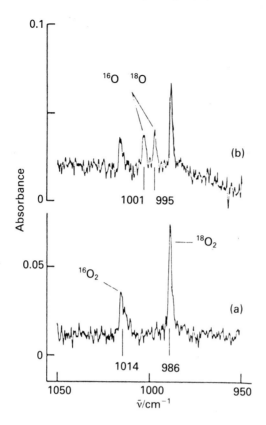

Fig. 6.2 — Infrared spectra recorded after photolysis of diazocyclopentadiene in $O_2$-doped $N_2$ matrices: (a) only $^{16}O_2$ and $^{18}O_2$ present; (b) $^{16}O_2$, $^{16}O^{18}O$, and $^{18}O_2$ present. Reproduced with permission from Bell et al., J. Chem. Soc., Chem. Commun., 1213 © 1983 Royal Society of Chemistry.

a 10% $O_2$-doped $N_2$ matrix. It is assumed that the reaction proceeds via the carbene, cyclopentadienylidene (**5**), evidence for this assumption being that a significant yield of 5 is observed alongside **6** when **4** is photolysed in an $N_2$ matrix containing only 1% $O_2$. Fig. 6.2 shows the $v(O-O)$ stretching region of the IR spectrum of various $^{18}O$ isotopomers of **6**.

**(4)** $\xrightarrow[\text{20 K}]{\lambda=300\text{ nm}}$ **(5)** $\xrightarrow{+ O_2}$ **(6)**

Around this time, Chapman and Hess carried out an independent study of the same reaction [8]. However, rather than irradiate their matrix with wavelengths of light close to 300 nm, as Bell and Dunkin had done, these workers used broad-band visible irradiation ($\lambda > 418$ nm). They observed a different oxidation product with two equivalent oxygen atoms. Dunkin and Shields later rationalized these observations [9], and proposed the following scheme for the photo-oxidation of cyclopentadienylidene where **8** represents the intermediate first observed by Chapman and Hess, and **6** the intermediate first observed by Bell and Dunkin.

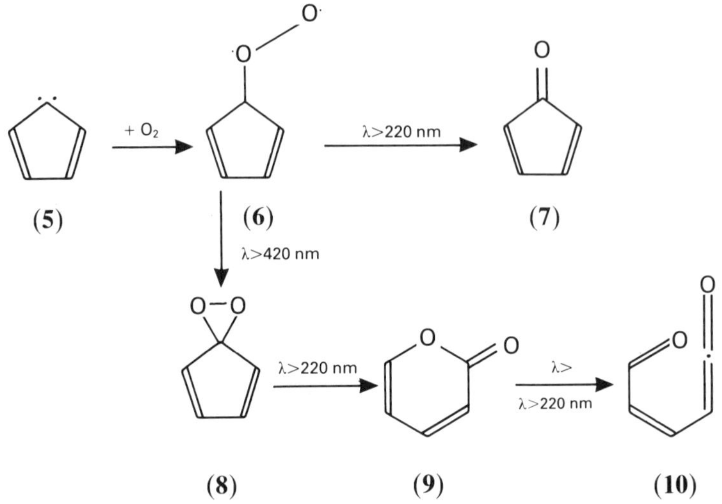

In this way the complex mechanism of the matrix photo-oxidation of cyclopentadienylidene, **5**, can be understood. One of the remarkable features of this reaction is the similarity between the intermediates **6** and **8** and the superoxo and peroxo derivatives of metal carbonyl fragments described in Chapter 3 and 4. Again the relationship between metal carbonyl fragments and organic carbenes is emphasized.

### 6.3  DECARBONYLATION REACTIONS

Just as carbenes will add to CO to form carbonyl compounds, photochemical loss of CO from a carbonyl precursor will yield a carbene. The most simple example of such a reaction is the flash photolysis of ketene (**11**) to yield methylene. In this way the UV-visible spectrum of gas phase methylene has been recorded, and the kinetics of its reactions monitored.

$$CH_2CO \xrightarrow{\ h\nu\ } CH_2 + CO$$
$$(11)$$

More complex carbonyl precursors yield carbenes, which may themselves undergo a range of rearrangement and further elimination reactions. Thus photolysis of matrix-isolated benzocyclobutenedione (**12**) results in a photo-equilibrium being attained with the *bis*-ketene (**13**). Continued photolysis slowly yields benzyne (**16**) by loss of two CO groups, a reaction which proceeds via the carbene **14** [10]. It appears that this carbene rearranges to yield the intermediate benzocyclopropenone (**15**), from which loss of CO gives the product benzyne. Indirect evidence for the

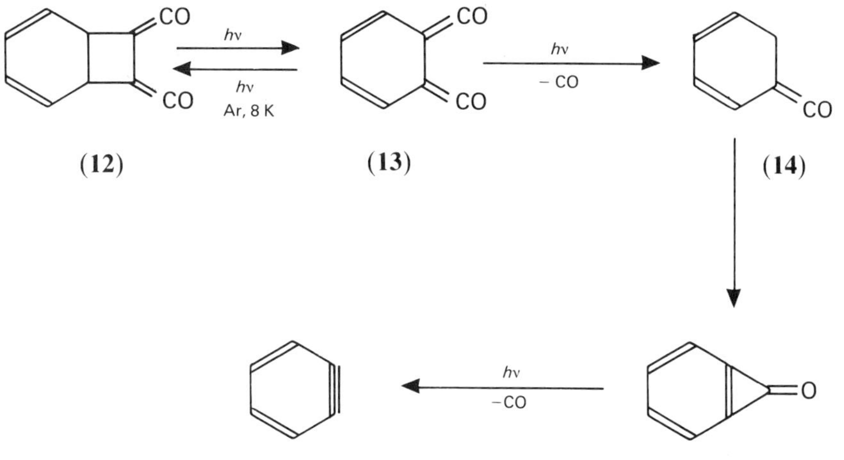

(12)                          (13)                          (14)

(16)                          (15)

intermediacy of **13** has come from experiments in which the process is carried out in a matrix containing methanol. On annealing, the dimethyl ester of *o*-phthalic acid (**17**) is formed, presumably by the reaction

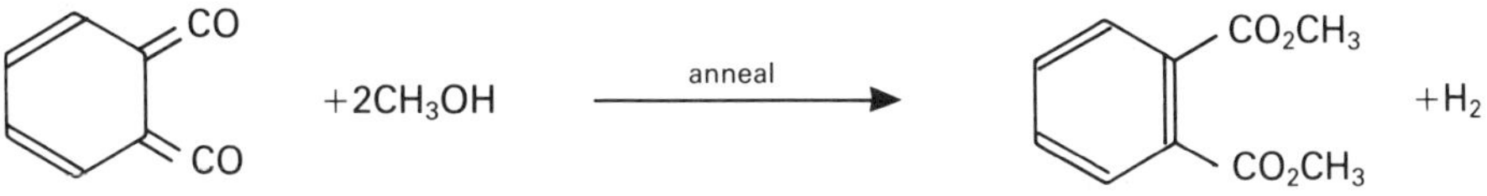

(13)                          (17)

These decarbonylation reactions, which yield carbenes, bear an obvious similarity to the photochemical reactions of metal carbonyls, in which CO is lost from the metal centre, and unsaturated metal carbonyl fragments are produced.

## 6.4   CYCLOBUTADIENE AND THE SEARCH FOR TETRAHEDRANE

The isolobal relationship may be extended to some quite complicated molecular species. Thus the methine group, CH, is related not only to the $d^9$ metal carbonyl fragment, $Ir(CO)_3$, but also to the $d^5$ cyclopentadiene derivative $CpW(CO)_2$ [4].

$$CH \longleftarrow \bigcirc \longrightarrow CpW(CO)_2 \longleftarrow \bigcirc \longrightarrow Ir\,(CO)_3$$

These small fragments may be 'joined together' to form organometallic clusters. For example, the tetrahedral cluster $Ir_4(CO)_{12}$ (**18**) may be thought of as a tetramer of the $Ir(CO)_3$ fragment, while the recently synthesized compound **19** is related to the well-known (cyclobutadiene)$Fe(CO)_3$ (**20**). The $W_2C_2$ ring of **19** corresponds to

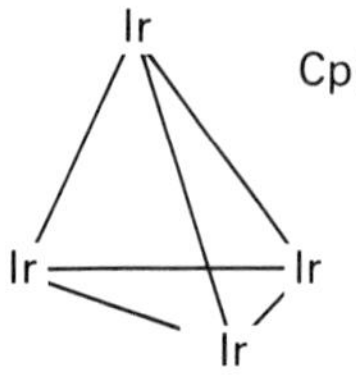

**(18)** 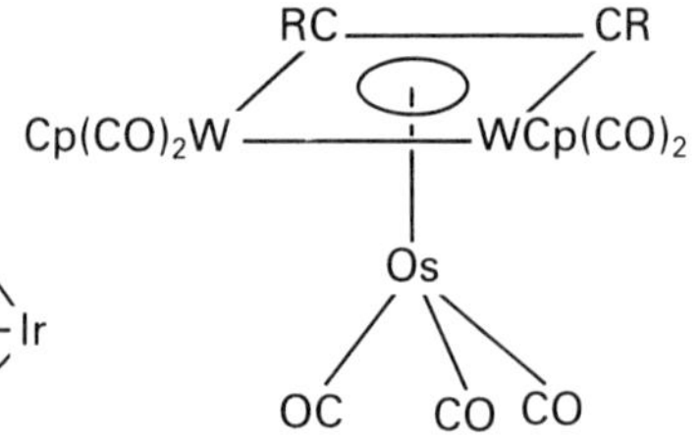        **(19)** 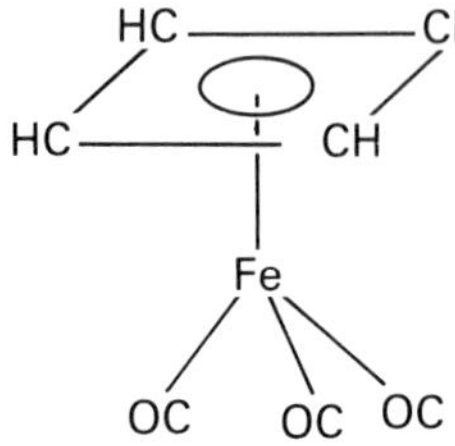        **(20)**

the $C_4$ ring of cyclobutadiene. Consequently, there is interest in studying the two organic isomers, cyclobutadiene **(21)** and tetrahedrane **(22)**, which provide simple analogues of these organometallic compounds.

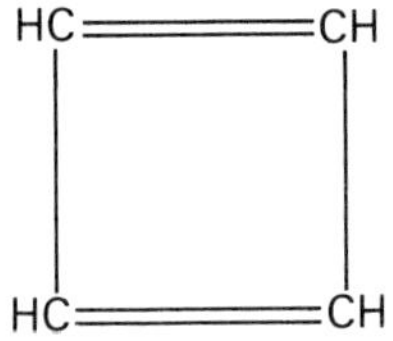

**(21)** 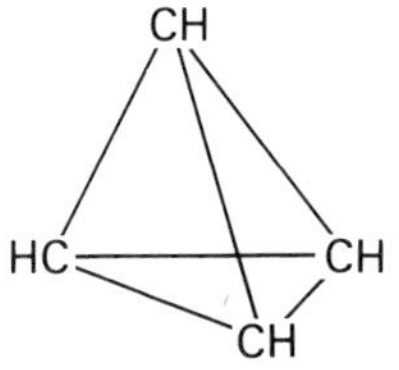                **(22)**

Cyclobutadiene has been the subject of several matrix-isolation experiments. The molecule was first prepared under matrix conditions by the groups of Krantz [11] and Chapman [12]. These workers independently used the same route to cyclobutadiene, i.e. the photolysis of α-pyrone **(23)** in an argon matrix at 8–20 K. The reactions proceed via the β-lactone **(24)** which loses $CO_2$ to yield cyclobutadiene. On further broad-band irradiation, cyclobutadiene is photolysed to two molecules of acetylene, while on annealing the matrix to 35 K the dimer of cyclobutadiene **(25)** is formed.

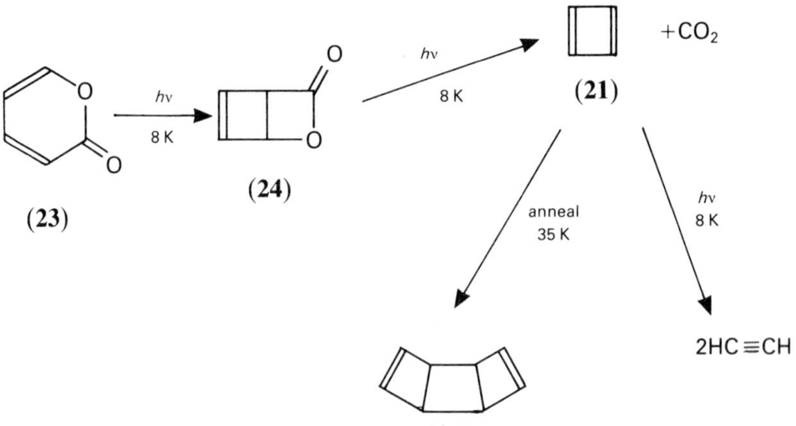

**(23)**        **(24)**        **(21)**    $+CO_2$

$2HC \equiv CH$

**(25)**

It is of importance to ascertain whether cyclobutadiene is a rectangular molecule of $D_{2h}$ symmetry or a square planar molecule of $D_{4h}$ symmetry. If its symmetry is $D_{4h}$, cyclobutadiene is expected to show only four infrared-active vibrations, whereas seven infrared-active vibrations are expected for the molecule if it has $D_{2h}$ symmetry. Chapman and his co-workers observed four infrared-active modes, and assigned them as shown in Table 6.1. Therefore, they arrived at the tentative conclusion that cyclobutadiene is a square planar molecule, a supposition which is

**Table 6.1** — Infrared absorptions of matrix-isolated cyclobutadiene

| $\nu/cm^{-1}$ | Assignment |
|---|---|
| 673 | Out-of-plane bend |
| 1037 | In-plane bend |
| 1482 | C−C stretch |
| 3064 | C−H stretch |

borne out by the results of isotopic substitution experiments [13]. An additional band, at 653 cm$^{-1}$, originally assigned to cyclobutadiene [11] appears instead to arise from a photo-ejected $CO_2$ molecule, which is perturbed by loose co-ordination to cyclobutadiene [14]. The band at 653 cm$^{-1}$ is not observed in the infrared spectrum of cyclobutadiene formed by routes — for example, pyrolysis of the iodinated hydrocarbons **26** and **27** — which do not involve production of $CO_2$ [15].

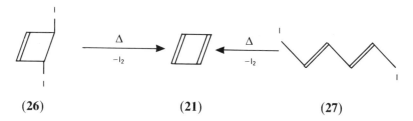

(26)                    (21)                    (27)

There has been some effort devoted by Günther Maier and his co-workers at Giessen to a search for the hydrocarbon tetrahedrane and its alkyl-substituted derivatives. Their approach has centred on photolysing the bicyclobutane dicarboxylic anhydrides **28**. Photolysis of the hydrogen- or methyl-substituted versions of **28** in low-temperature argon matrices yields only the corresponding cyclobutadiene. It appears that the lactone **24** is an intermediate in the reaction [16].

(28)                    (24)                    (21)

However, the tetra-*tert*-butyl derivative of tetrahedrane (**29**) has been prepared by photolysis of tetra-*tert*-butylcyclopentadienone (**30**). This synthesis [17] provides the first example of the formation of an alkyl-substituted tetrahedrane. The product exists as colourless stable crystals; it isomerizes to the corresponding cyclobutadiene (**31**) on heating to 135°C, but the reaction may be reversed photochemically.

(**30**)

(**29**)

(**31**)

## 6.5  SUMMARY

This short chapter mentions only a very small part of the work which has been carried out on unstable organic molecules. However, the results described indicate some of the analogies which may be drawn between the chemistry shown by short-lived inorganic and organic species. Perhaps the most striking examples of this similarity are the reactions of organic carbenes and of unsaturated metal carbonyl fragments — species which are related by the isolobal relationship. Thus the elegant work of Dunkin and Chapman on organic oxides formed by the reaction of carbenes with dioxygen appears to be closely related to the work carried out on photo-oxidation reactions of metal carbonyls in low-temperature matrices, which was described in Chapter 3. Conventional ideas of what constitutes a 'complex' appear to be somewhat misleading here. The importance of characterizing short-lived intermediates is emphasized. It is only by identifying these unstable species — in particular using matrix isolation — that the mechanisms of the photo-oxidation of carbenes and of metal carbonyls have become apparent. In this way new insights into the bonding of oxygen to unsaturated inorganic and organic species have been gained. These experimental results provide an intriguing example of the isolobal relationship whereby species, which at first sight seem to bear little resemblance to each other, show some remarkably similar chemical reactions.

That the reactions of carbenes and silylenes are related is not surprising. However, it is interesting to compare the chemistry of the different electronic states of these molecules. Thus *ground-state* triplet silylenes show reactions — e.g. stereospecific addition to C=C double bonds — which are also shown by *excited-state* triplet carbenes. The reactivity of ground-state singlet carbenes is somewhat different. Although, for example, these species will add to C=C double bonds, they do so in a non-stereospecific manner. Because of their different electronic states it is seen that, in general, ground-state singlet carbenes are more reactive but less selective than their triplet silylene counterparts. Other differences include dimerization reactions, which are quite well known for silylenes but are rarely met in carbene chemistry. Again it is only by careful study of these short-lived molecules that such comparisons have been possible.

# REFERENCES

[1] R. A. Bernheim, H. W. Bernard, P. S. Wang, L. S. Wood and P. S. Skell, *J. Chem. Phys.* (1970), **53**, 1280; (1971), **54**, 3223.

[2] J. F. Harrison and L. C. Allen, *J. Am. Chem. Soc.* (1969), **91**, 807; C. F. Bender and H. F. Schaefer III, *J. Am. Chem. Soc.* (1970), **92**, 4984.

[3] L. Andrews and T. G. Carver, *J. Chem. Phys.* (1968), **49**, 896.

[4] R. Hoffman, *Angew. Chem. Int. Ed. Engl.* (1982), **21**, 711.

[5] M. Poliakoff, *Chem. Soc. Rev.* (1978), **7**, 527; M. J. Almond, A. J. Downs and M. Fanfarillo, unpublished results; M. Fanfarillo, D. Phil. Thesis, University of Oxford (1988).

[6] M. S. Baird, I. R. Dunkin, N. Hacker, M. Poliakoff and J. J. Turner, *J. Am. Chem. Soc.* (1981), **103**, 5190.

[7] G. A. Bell and I. R. Dunkin, *J. Chem. Soc., Chem. Commun.* (1983), 1213.

[8] O. L. Chapman and T. C. Hess, *J. Am. Chem. Soc.* (1984), **106**, 1842.

[9] I. R. Dunkin and C. J. Shields, *J. Chem. Soc., Chem. Commun.* (1986), 154.

[10] O. L. Chapman, *Pure & Appl. Chem.* (1974), **40**, 511.

[11] C. Y. Lin and A. Krantz, *J. Chem. Soc., Chem. Commun.* (1972), 1111.

[12] O. L. Chapman, C. L. McIntosh and J. Pacansky, *J. Am. Chem. Soc.* (1973), **95**; 617.

[13] O. L. Chapman, D. De La Cruz, R. Roth and J. Pacansky, *J. Am. Chem. Soc.* (1973), **95**, 1337.

[14] R. G. S. Pong, B. S. Huang, J. Laureni and A. Krantz, *J. Am. Chem. Soc.* (1977), **99**, 4153.

[15] G. Maier, M. Hoppe, K. Lanz and H. P. Reisenauer, *Tetrahedron Lett.* (1984), **25**, 5645.

[16] H. W. Lage, H. P. Reisenauer and G. Maier, *Tetrahedron Lett.* (1982), **23**, 3893; G. Maier and H. P. Reisenauer, *Chem. Ber.* (1981), **114**, 3959.

[17] G. Maier, S. Pfriem, U. Schäfer, K. D. Malsch and R. Matusch, *Chem. Ber.* (1981), **114**, 3965.

# 7

# High-temperature molecules

## 7.1 INTRODUCTION

Many chemical molecules, which are not stable under normal conditions at room temperature, are found in high-temperature vapours. These include vapour phase molecular forms of compounds such as metal oxides and chlorides, which exist as solids at room temperature. Other interesting species found in high-temperature vapours have unusual valencies and co-ordination numbers. Examples are the molecules OPCl and $O_2PCl$ which contain unusual two-co-ordinate trivalent and three-co-ordinate pentavalent phosphorus atoms respectively.

Some methods of characterization of high-temperature molecules rely on direct vapour phase measurements, e.g. mass spectrometric and electron diffraction studies. Matrix isolation has, however, proved to be a good method for trapping the species present in high-temperature vapours, and in this way many spectroscopic studies have been made. Perhaps the most unambiguous conclusions have been drawn from a combination of vapour phase and matrix-isolation studies.

A recent development in this field has been the use of high-temperature molecules in synthetic chemistry. These experiments are similar to metal vapour synthesis, which is described in section 4.11. Here, however, rather than metal atoms, small molecules are condensed with suitable ligands at liquid nitrogen temperature. It is hoped that many new and interesting molecules will be prepared in this way.

## 7.2 SILICON MONOXIDE

Thermodynamic calculations have shown that the principal vapour phase species in the silicon/oxygen system is the monoxide molecule SiO. In 1969, Anderson and Ogden published the results of matrix-isolation experiments in which the vapour over heated silica (at 2000 K) or over heated silicon/silica mixtures (at 1600 K) was co-condensed with excess argon or nitrogen [1]. These results show that alongside the monomer SiO, the dimer $Si_2O_2$ (1) and trimer $Si_3O_3$ (2) are also trapped in the matrix. Fig. 7.1 illustrates the infrared spectrum of silicon oxides prepared under

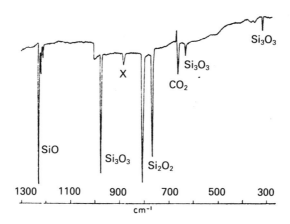

Fig. 7.1 — Infrared spectrum of silicon oxides in a nitrogen matrix. Reproduced with permission from Anderson and Ogden, *J. Chem. Phys.*, **51**, 4189 © 1969 American Institute of Physics.

these conditions and isolated in a nitrogen matrix. The assignment of the bands to the molecules SiO, $Si_2O_2$ and $Si_3O_3$ is given. One of the problems inherent in interpreting matrix-isolation experiments on high-temperature molecules is to determine to what extent the matrix-isolated species reflect the composition of the high-temperature vapour. Mass spectrometric studies have indicated that $Si_2O_2$ but not $Si_3O_3$ is present alongside SiO in the vapour over heated silicon/silica mixtures [2]. The presence of $Si_3O_3$ in the matrix-isolated samples suggests that this molecule is formed during the freezing of the matrix, and indicates that $Si_3O_3$ might be an important intermediate in the polymerization of SiO.

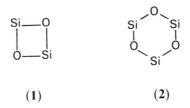

**(1)**                    **(2)**

Such aggregation processes are, however, not well understood. Silicon monosulphide, SiS — prepared by passing $CS_2$ gas over silicon heated to 1270 K — and the isoelectronic molecule PN — prepared by heating the solid $P_3N_5$ to 1070–1170 K — have both been the subjects of similar matrix-isolation experiments [3]. $Si_2S_2$ but no higher polymers are seen alongside SiS in the matrix, but PN trimerizes to $P_3N_3$, a cyclic molecule, isostructural with $Si_3O_3$. Fig. 7.2 shows the buildup of the infrared bands associated with the $^{14}N/^{15}N$ isotopomers of $P_3N_3$ and the concomitant decay of the bands of an equimolar mixture of $P^{14}N$ and $P^{15}N$ when a krypton matrix is annealed to 35 K. There is no sign of a dimeric molecule, $P_2N_2$, in any of these experiments.

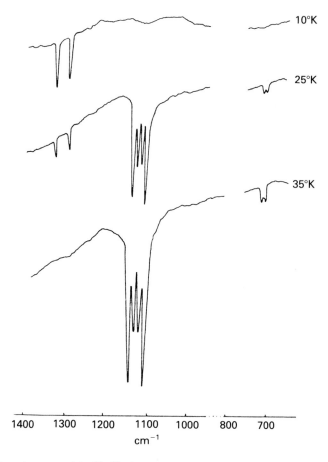

Fig. 7.2 — Infrared spectra of the $^{14}N/^{15}N$ isotopomers of PN and $P_3N_3$ in a krypton matrix. Reproduced with permission from Atkins *et al.*, *Spectrochim. Acta part A*, **33A**, 853 © 1977 Pergamon.

The chemical reactions of matrix-isolated SiO are of much interest, in part because of their relationship to the reactions of other divalent silicon species, which are described in Chapter 5. As expected, both addition and insertion reactions are observed.

Matrix-isolated silicon dioxide, $SiO_2$, was first prepared by Schnöckel, who co-condensed vapour phase SiO with atomic oxygen (generated by microwave excitation of $O_2$) and excess argon. Alongside the absorptions of $O_3$ and unreacted SiO, the infrared spectrum of such a matrix shows a band of medium intensity at 1416.5 cm$^{-1}$, which is assigned to the antisymmetric stretch of a linear $SiO_2$ molecule [4,5]. Isotopic substitution experiments with $^{18}O$ confirm the linearity of this molecule. Force constant calculations show that the Si–O bonds of $SiO_2$ are rather weaker than the C–O bonds of $CO_2$, presumably as a result of the less efficient overlap of orbitals on atoms from the first and second periods. It is also noteworthy that the carbonyl sulphide analogue, OSiS, has been generated by the matrix reaction of SiS with O atoms [6]. This molecule is of interest because it contains both Si=O and Si=S double bonds.

Irradiation of an argon matrix containing SiO and $Cl_2$ leads, via an insertion reaction, to the product $OSiCl_2$ — the silicon analogue of phosgene [5]. Its infrared spectrum shows this to be a planar molecule with a force constant $f(SiO)$ very similar to that calculated for $SiO_2$.

$$SiO + Cl_2 \xrightarrow{h\nu} OSiCl_2$$

Recently, experiments have been performed to ascertain if SiO, like CO, will co-ordinate to transition metals. Co-condensation of silver atoms with monomeric SiO and an excess of argon yields Ag(SiO) and a second molecule which is probably $Ag_2(SiO)$ [7]. Isotopic studies involving the natural abundance of $^{28}Si$, $^{29}Si$ and $^{30}Si$ (see Fig. 7.3) and substitution with $^{18}O$ imply that the molecule Ag(SiO) has a

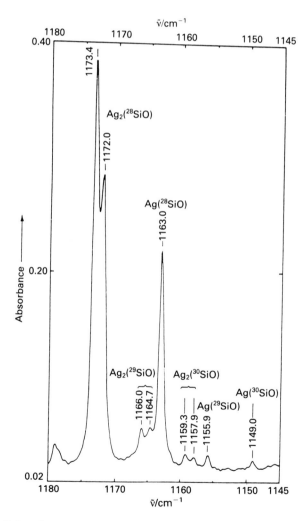

Fig. 7.3 — Infrared spectra of the products of the reaction of SiO and Ag in an argon matrix. Reproduced with permission from Mehner *et al.*, *J. Chem. Soc., Chem. Commun.*, 117 © 1988 Royal Society of Chemistry.

triangular arrangement with side-on bonding of SiO. Two possible structures, which cannot easily be distinguished by the isotopic substitution studies, are **3** and **4**. It is

Ag——Si                                    Si
      \|                           Ag<   |
       O                                  O

(3)                                    (4)

clear, however, that the molecule does not have the linear structure (**5**) characteristic of matrix-isolated metal monocarbonyls, e.g. Ni(CO) and Rh(CO). By contrast it seems that the molecule Au(SiO), formed by co-condensation of gold atoms and SiO with excess argon, adopts the linear structure **6**. It appears that the mode of co-ordination of SiO to a metal is affected much more by the nature of the metal than is the mode of co-ordination of CO. The results on the systems Ag/Si/O and Au/Si/O may have important implications for research and development of silicon-based electronic devices, which are often made conducting by partially coating with gold.

Ni–C–O                              Au–Si–O

(5)                                    (6)

### 7.3 SMALL ALUMINIUM MOLECULES

Monovalent aluminium molecules bear some resemblance, electronically, to divalent silicon molecules. Examples of such molecules are the aluminium monohalides, which are found in the high-temperature vapours above a variety of chemical systems. For example, AlCl may be prepared by reacting aluminium vapour, at about 1000 K, with $AlCl_3$ vapour at about 350 K, while AlBr may be generated by a

$$AlCl_3 + 2Al \xrightarrow{\Delta} 3AlCl$$

number of reactions, including

$$2Al + MgBr_2 \xrightarrow{1250°C} 2AlBr + Mg$$

and

$$2Al + Br_2 \xrightarrow{1000°C} 2AlBr$$

Thus rotational spectra have been recorded for all four aluminium halides in the gas phase [8]. From these spectra, rotational and vibrational constants have been calculated. It is also possible to trap aluminium monohalide molecules in low-temperature matrices by co-condensing the vapour with an excess of inert gas [9]. In this way, infrared spectra have been recorded. Fig. 7.4 shows the infrared spectrum of matrix-isolated AlCl; the absorption is clearly resolved into two components corresponding to the natural abundance of $^{35}Cl$ and $^{37}Cl$. Table 7.1 compares the

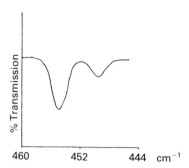

Fig. 7.4 — Infrared spectrum of AlCl in an argon matrix. Reproduced with permission from Schnöckel, *Z. Naturforsch B*, **31B**, 1291 © 1976 Verlag der Zeitschrift fur Naturforschung.

**Table 7.1** — Wavenumbers (cm$^{-1}$) of the vibrations of gas phase and matrix-isolated AlCl and AlBr

|                | Gas phase | Ne matrix | Ar matrix | N$_2$ matrix |
|----------------|-----------|-----------|-----------|--------------|
| Al$^{35}$Cl    | 477.5     | 470       | 455       | —            |
| Al$^{79}$Br    | 374.4     | 370       | 357       | 349          |

frequencies of the vibrations of gas phase and matrix-isolated AlCl and AlBr. These values indicate that the molecules when trapped in a neon matrix most closely resemble those in the gas phase.

Some interesting chemical reactions have been observed for matrix-isolated aluminium monohalides. AlF, prepared by reaction of Al with HF at 1100 K, will dimerize in rare gas matrices to form Al$_2$F$_2$ which has a cyclic structure **7** [10]. In argon matrices containing oxygen atoms, the molecule OAlF (**8**) is produced [11], while in dioxygen-doped matrices, the interesting planar molecule (**9**), which has $D_{2h}$ symmetry, is generated [10]. The structures of these molecules have been determined not only by isotopic substitution experiments with $^{18}$O, but also by *ab initio* calculations. This work emphasizes the importance of combining experimental measurements with theoretical calculations.

$$
\begin{array}{ccc}
\underset{\textbf{(7)}}{\underset{\text{F}}{\overset{\text{F}}{\text{Al}\diamondsuit\text{Al}}}} &
\underset{\textbf{(8)}}{\text{O}=\text{Al}-\text{F}} &
\underset{\textbf{(9)}}{\text{F}-\text{Al}\overset{\text{O}}{\underset{\text{O}}{\diamondsuit}}\text{Al}-\text{F}}
\end{array}
$$

AlCl will react, in argon matrices, with oxygen atoms to form OAlCl, an analogue of OAlF [10], or with HCl to form the planar molecule HAlCl$_2$ (**10**) [11].

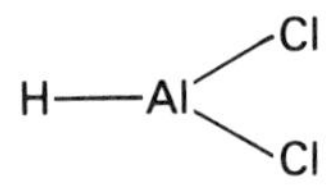

$$
H\!-\!Al\!\!\begin{array}{c} \diagup Cl \\ \diagdown Cl \end{array}
$$

**(10)**

The insertion reaction of AlCl into the HCl bond is reminiscent of reactions undergone by divalent silicon molecules (see Chapter 5). The compound SAlCl may be prepared as a stable solid by the reaction

$$AlCl_3 + Al_2S_3 \xrightarrow{\sim 600\ K} 3SAlCl$$

The structure of the solid has been investigated by infrared and Raman spectroscopy and by diffraction techniques [12]. It appears to consist of chains containing $(SAlCl)_2$ units **(11)**, a structure which is related to that seen for matrix-isolated $(OAlF)_2$ molecules **(9)**.

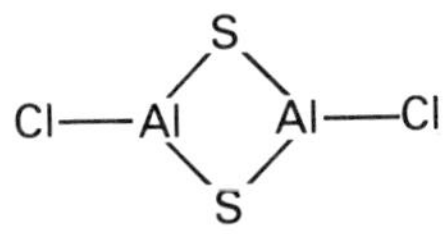

$$
Cl\!-\!Al\!\!\begin{array}{c} \diagup S \diagdown \\ \diagdown S \diagup \end{array}\!\!Al\!-\!Cl
$$

**(11)**

The monomer AlCl may be employed in preparative chemical reactions. Co-condensation of AlCl vapour with dimethylacetylene at liquid nitrogen temperature (77 K) leads, on warming the reaction mixture to room temperature, to the formation of an air-and-moisture-sensitive product [13]. The chemical composition and structure of this product have been investigated by elemental analysis (showing an Al:Cl ratio of 1:1), $^1$H NMR (showing a singlet peak at $\delta = 2.08$, corresponding to methyl group resonance), mass spectrometry (suggesting a molecular formula of $Al_4Cl_4C_{16}H_{24}$) and single crystal diffraction studies at $-130°C$. The diffraction studies show that the molecule is a dimeric 1,4-dialumina-2,5-cyclohexadiene **(12)**. The two rings of the dimer are joined by a $\pi$ interaction between the Al atoms of one ring and the C=C bonds of the other **(13)**. The C=C bond is somewhat longer (136.7 pm) than an unco-ordinated C=C bond, while the mean Al–C distance is

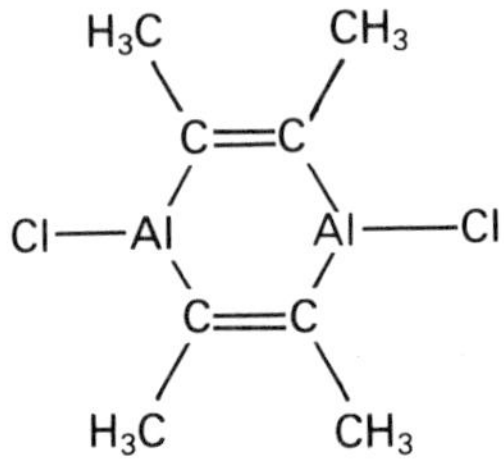

**(12)**

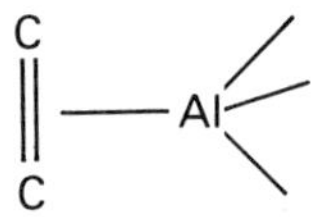

**(13)**

235.4 pm, i.e. approximately 10% greater than the Pt–C distance in the well-known alkene complex, Zeise's salt, $K[Pt(C_2H_4)Cl_3]$. This molecule is the first example of a compound containing the structural unit **13** and is of interest because it may be similar to intermediates in reactions of organoaluminium compounds, perhaps even in polymerization reactions catalyzed by Ziegler–Natta catalysts. It is hoped that similar reactions may lead to the preparation of many new compounds from the short-lived molecules present in high-temperature vapours.

## 7.4  SMALL PHOSPHORUS MOLECULES

Small phosphorus molecules with unusual valencies and co-ordination numbers are found in high-temperature vapours. An example of such a molecule is the phosphaethene $CH_2=PCl$, which contains a trivalent, two-co-ordinate phosphorus atom. This species has been prepared by pyrolysis of methyldichlorophosphite,

$$CH_3OPCl_2 \xrightarrow{\Delta} CH_2=PCl + HOCl$$

$CH_3OPCl_2$, and has been studied by microwave spectroscopy [14]. Thus its structure has been determined. It is shown to have a C=P distance close to 165.5 pm, a P–Cl distance close to 206.0 pm and a CPCl angle of 103.3° (**14**).

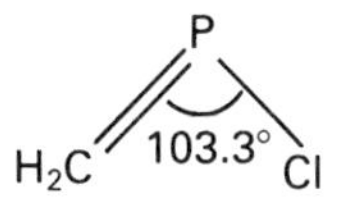

(**14**)

Isoelectronic with $CH_2=PCl$ is the molecule OPCl. This molecule has been prepared by two separate routes:

$$POCl_{3(g)} + Ag_{(s)} \xrightarrow{1100\ K} OPCl_{(g)} + 2AgCl_{(g)}$$

and

$$PCl_{3(g)} + H_2O_{(g)} \xrightarrow{700\ K} OPCl_{(g)} + 2HCl_{(g)}$$

and has been studied both by infrared spectroscopy of the matrix-isolated species, and by mass spectrometric studies on the vapour [15]. As shown in Fig. 7.5, all three fundamental vibrations of OPCl have been observed in the infrared spectrum. Measurement of isotopic shifts arising from the natural abundance of $^{37}Cl$, or from substitution with $^{18}O$ has allowed the OPCl angle to be calculated at ca. 105°, i.e. close to the bond angle of $CH_2PCl$.

It is possible to estimate the standard enthalpy of formation of OPCl [$\Delta H^0_{f298}$ (OPCl$_g$)] from the mass spectrometric data. The measured appearance potential of

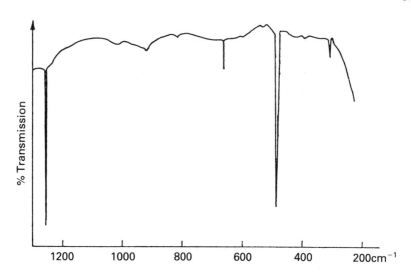

Fig. 7.5 — Infrared spectrum of OPCl in an argon matrix. Reproduced with permission from Binnewies *et al.*, *Z. Anorg. Allg. Chem.*, **497**, 7 © 1983 Johann Ambrosius Barth.

$P^+$ [$AP(P^+)$] in the mass spectrum of OPCl is 20.9 eV. From the Born–Haber cycle given below it is apparent that the enthalpy of atomization of OPCl [$\Delta H^0_{at}(OPCl)$] can be estimated from

$$\Delta H^0_{at}(OPCl) = AP(P^+) - I(P)$$

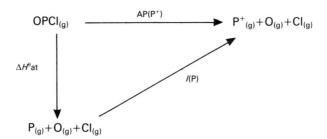

Born–Haber cycle

where $I(P)$ is the first ionization potential of phosphorus. Thus $\Delta H^0_{at}(OPCl)$ is calculated as 9.9 eV or 955 kJ mol$^{-1}$. The standard enthalpies of formation of $P_{(g)}$ (333.9 kJ mol$^{-1}$), $O_{(g)}$ (249.4 kJ mol$^{-1}$) and $Cl_{(g)}$ (121.0 kJ mol$^{-1}$) are all known, so $\Delta H^0_{f298}$ (OPCl) can thus be calculated as − 250.7 kJ mol$^{-1}$.

Other molecules, similar to OPCl, have been generated in high-temperature vapours trapped in low-temperature matrices and studied by infrared spectroscopy. These include SPCl [16] and OPF [17], prepared by the reactions

$$SPCl_{3(g)} + 2Ag_{(s)} \xrightleftharpoons{1100 \text{ K}} SPCl_{(g)} + 2AgCl_{(g)}$$

and

$$OPFBr_{2(g)} + 2Ag_{(s)} \xrightleftharpoons{1000 \text{ K}} OPF_{(g)} + 2AgBr_{(g)}$$

For both molecules, the bond angle at the phosphorus atom is about 110°. OPF has also been the subject of mass spectrometric investigations and elaborate *ab initio* calculations [17].

Even more uncommon than molecules containing trivalent, two-co-ordinate phosphorus atoms are those which contain a pentavalent, three-co-ordinate phosphorus atom. An example of such a molecule is $O_2PCl$ [18], which may be prepared either by a photochemical reaction between OPCl and $O_3$ in an argon matrix:

$$OPCl + O_3 \xrightarrow[\text{Ar matrix}]{h\nu} O_2PCl + O_2$$

or in the high-temperature gas phase equilibrium:

$$OPCl_{3(g)} + \tfrac{1}{2}O_{2(g)} + 2Ag_{(c)} \xrightleftharpoons{1300 \text{ K}} 2AgCl_{(g)} + O_2PCl_{(g)}$$

The infrared spectra of matrix-isolated $O_2PCl$ show that it is a planar molecule with an $O-\hat{P}-O$ bond angle of about 135° (**15**) and an unusually strong P–Cl bond. These results have been confirmed by *ab initio* calculations, and mass spectrometric measurements have also been made on the vapour phase species.

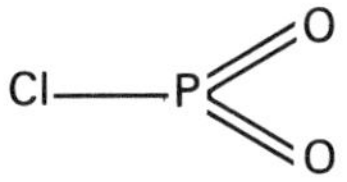

Cl——P$\diagdown$ $^O_O$

(**15**)

## 7.5  METAL OXYSALTS
Many structural studies have been made on the species present in the vapours from oxysalts such as alkali metal nitrates, chlorates, perrhenates, etc. In part, the interest in such work lies in determining the mode of co-ordination of the metal cation to the oxyanion in the monomeric vapour phase species. Two experimental approaches which have been employed are vapour phase electron diffraction and matrix-isolation infrared spectroscopy.

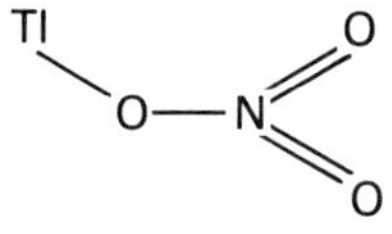

(**16**) 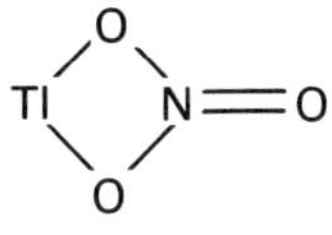        (**17**) 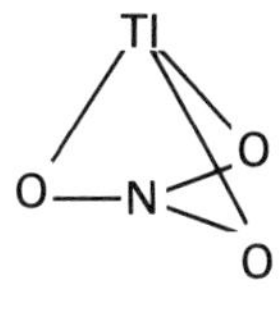        (**18**)

Three plausible structures may be envisaged (**16,17,18**) for a simple monomeric metal nitrate such as $TlNO_3$. Electron diffraction studies [19] have indicated that the bidentate structure (**17**) is preferred. Likewise thallium perrhenate ($TlReO_4$) shows bidentate co-ordination of the $Tl^+$ cation (**19**) to the distorted tetrahedral $ReO_4^-$ anion [20].

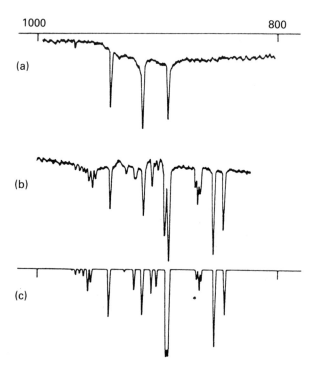

(**19**)

A bidentate structure, analogous to that determined by electron diffraction measurements for $TlReO_4$, has been observed for matrix-isolated $MReO_4$ and $MClO_4$ molecules, where M = Li, K or Cs. Here the structure is deduced by means of infrared and Raman spectroscopy including isotopic substitution with $^{18}O$ [21,22]. Fig. 7.6 shows the observed and calculated infrared spectra for $CsReO_4$ 50% enriched with $^{18}O$, assuming bidentate co-ordination, i.e. $C_{2v}$ molecular symmetry;

Fig. 7.6 — Infrared spectra of Cs $ReO_4$ in nitrogen matrices: (a) Cs $Re^{16}O_4$; (b) $CsReO_4$ 50% enriched in $^{18}O$; (c) calculated spectrum assuming a bidentate structure. Reproduced with permission from Arthers *et al.*, *J. Chem. Soc., Dalton Trans.*, 1461 © 1983 Royal Society of Chemistry.

it can be seen that a very good correlation is obtained. The results of these experiments seem to have settled finally an earlier inconsistent electron diffraction study [23] on $KReO_4$, which was interpreted assuming a monodentate co-ordination. That $KReO_4$, like $TlReO_4$ and $CsReO_4$, has a bidentate co-ordination is now certain.

Other metal oxysalts which show bidentate co-ordination are metal phosphates $(MPO_3)$ and phosphites $(MPO_2)$, and their arsenic analogues (**20,21**) [24,25]. The generation of molecular $NaPO_2$ is of interest. This molecule is not produced by direct evaporation of a salt; instead it is formed by reduction of $NaPO_3$ by a reducing metal such as molybdenum or tantalum, incorporated in the cell from which the sample is vaporized [24].

(**20**)                    (**21**)            (E = P or As)

In contrast to the above molecules, caesium chlorate, $CsClO_3$, provides a rare example of tridentate co-ordination of a metal cation to an oxyanion (**22**). Measurement of the IR spectra of the various $^{18}O$ isotopomers of this molecule, when trapped in an argon matrix, has allowed the $C_{3v}$ structure (**22**) to be clearly distinguished from any other possible molecular geometry [26].

(**22**)

Some interesting results have been reported by Devlin [27] regarding the *solvation* of $LiNO_3$ molecules in low-temperature matrices. In a pure argon matrix, $LiNO_3$ adopts structure **17** and the degenerate $v_3$ mode of the trigonal nitrate ion is split into two components separated by 262 cm$^{-1}$. However, if $LiNO_3$ is condensed with argon containing Lewis bases such as water or tetrahydrofuran (THF), then the Lewis base molecules co-ordinate to the lithium atom. The splitting of $v_3$ thus decreases in regular steps corresponding to the stepwise increase in co-ordination number (CN) of the lithium atom. Fig. 7.7 illustrates the infrared spectra of $LiNO_3$ isolated in argon matrices doped with concentrations of THF ranging from 1 to 25%, and Table 7.2 lists the frequencies of matrix-isolated $LiNO_3$ solvated by different numbers of either THF or water molecules. One remarkable feature of these experiments is that in general the co-ordination number of the lithium atom reaches a higher value in the matrix than it does in solution.

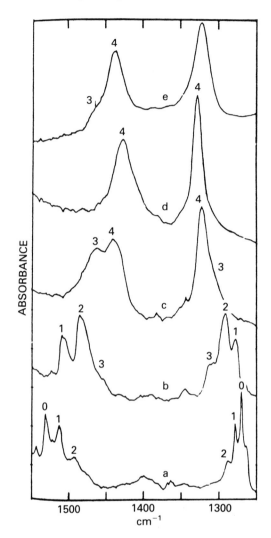

Fig. 7.7 — Infrared spectra of LiNO$_3$ in: (a) 1% THF-doped solid argon; (b) 7% THF-doped solid argon; (c) 25% THF-doped solid argon; (d) pure solid THF (all at 20 K); (e) pure THF solution at 298 K. The numbers refer to $n$ in LiNO$_3$.(THF)$_n$. Reproduced with permission from Ritzhaupt *et al.*, *J. Chem. Phys.*, **82**, 1167 © 1985 American Institute of Physics.

**Table 7.2** — Wavenumbers (cm$^{-1}$) of infrared absorptions of matrix-isolated LiNO$_3$ solvated by THF or water molecules

| CN solvent | 0 — | 1 H$_2$O | 1 THF | 2 H$_2$O | 2 THF | 3 H$_2$O | 3 THF | 4 H$_2$O | 4 THF |
|---|---|---|---|---|---|---|---|---|---|
| $\nu_{3a}$ | 1531 | 1520 | 1512 | 1500 | 1488 | 1480 | 1465 | 1437 | 1432 |
| $\nu_{3b}$ | 1269 | 1279 | 1279 | 1290 | 1290 | 1305 | 1315 | 1337 | 1327 |
| $\Delta\nu_3$ | 262 | 241 | 233 | 210 | 198 | 175 | 150 | 100 | 105 |

Some metal oxysalts have been prepared *in situ* in low-temperature matrices by co-condensing a metal oxide vapour with a reactive molecule and excess argon. Thus $Tl_2O$-generated in the gas phase by direct vaporization of a solid sample reacts with $SO_2$ to yield $Tl_2SO_3$ or $Tl_2S_2O_5$ molecules, or with OCS to yield $Tl_2CO_2S$ [28]. The structures of these products have been investigated by infrared spectroscopy.

$$Tl_2O + SO_2 \xrightarrow[\text{excess Ar}]{\text{co-condense}} Tl_2SO_3$$

## 7.6   METAL OXIDES AND HALIDES

The formation of matrix-isolated metal oxide molecules either by photolysis of metal carbonyls or by reactions of metal atoms is described in sections 3.9 and 4.3. However, a third route to such species is direct condensation of a vapour containing metal oxide molecules with an excess of inert gas.

In this way both molybdenum and tungsten oxides have been trapped in argon and neon matrices and it is interesting to compare the results of these experiments with those of the photolysis studies (see section 3.9). A mixture of $MoO_2$, $MoO_3$ and $(MoO_3)_{2-5}$ molecules is produced by the reaction, at 1700–2300°C, of molybdenum metal with $O_2$ [29]. Following condensation with excess argon or neon, the structures of these species have been deduced by infrared spectroscopy including isotopic substitution with $^{18}O$, and the observation of resolved separate infrared absorptions due to the different isotopes of molybdenum in their natural abundance. Thus it is shown that $MoO_2$ is a bent molecule with a bond angle of $\sim 118°$ (**23**), while $MoO_3$ is a trigonal planar molecule of $D_{3h}$ symmetry (**24**). Similar structures are found for matrix-isolated $WO_2$ and $WO_3$ produced, in this case, by sputtering a tungsten electrode in a mixture of $O_2$ and an inert gas [30].

(23)                                    (24)

These matrix studies have been complemented by electron diffraction measure-ments on the vapour present over solid $WO_3$ heated to 1250–1700 K in the presence or absence of water vapour [31]. Under these conditions the principal gas phase species appears to be the trioxide trimer, $W_3O_9$, and the most likely structure for this is a six-membered ring (**25**). Similarly, $W_3O_9$ has been detected, alongside $W_2O_6$, $W_4O_{12}$ and other species, in the mass spectrum of the vapour over a sample of $WO_3$ heated to ca. 1550 K [32].

Simple alkali metal halide molecules — for example NaCl and KCl — are found in the vapours above heated samples of the solid compounds. These molecules have

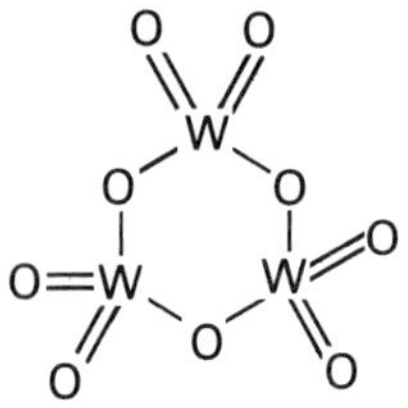

(25)

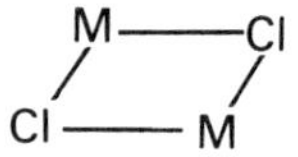

(26)

been studied by gas phase electron diffraction [33] and by infrared spectroscopy both on matrix-isolated [34] and on gas phase samples. Also present in such vapours are molecular dimers. These adopt a planar rhombic structure (26) in which the Cl–$\hat{\text{M}}$–Cl angle ($\theta$) decreases as the size of the alkali metal (M) increases. Values of $\theta$ for different alkali metal chloride dimers are given in Table 7.3.

**Table 7.3** — Cl–$\hat{\text{M}}$–Cl bond angles ($\theta$) for different alkali metal chloride dimers

| Metal | $\theta/°$ |
|-------|-----------|
| Na | 101 |
| K | 96 |
| Rb | 88 |
| Cs | 84 |

There has been some controversy regarding the structure adopted by the dihalide molecules of the first-row transition metals iron, cobalt and nickel. The infrared spectra of these molecules either in the gas phase [35] or when trapped in inert gas matrices [36] show no sign of the symmetric stretching vibration ($v_1$). Since this vibration is infrared inactive for a linear $MCl_2$ molecule a consensus of opinion was reached that these molecules have a linear equilibrium structure (27). This supposition was given support by the results of molecular beam deflection studies which show no sign of the permanent electric dipole moment expected for $MCl_2$ molecules with non-linear equilibrium geometry (28) [37]. Electron diffraction studies were more difficult to interpret. At first sight, the results of these experiments can be interpreted either in terms of a linear structure (27) or in terms of a slightly bent geometry with a bond angle of about 160° (28) [38]. A more recent matrix-isolation study by Green et al. has suggested that $FeCl_2$, $CoCl_2$ and $NiCl_2$ adopt structure 28 when trapped in argon matrices [39]. The situation has been clarified by Hargittai of

(27)

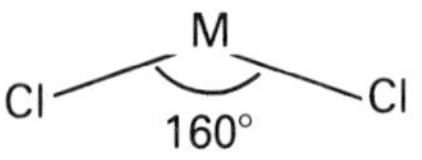

(28)

the Hungarian Academy of Sciences [40]. His careful interpretation of the electron diffraction data shows that the equilibrium geometry of the gas phase molecules is indeed linear, but that a matrix-induced structural change may account for the non-linear geometry observed by Green *et al*.

The structure of the aluminium trichloride monomer, $AlCl_3$, has been the subject of a similar controversy. Here, two matrix-isolation studies were in conflict. Lesiecki and Shirk preferred a pyramidal structure of $C_{3v}$ symmetry, and with a $Cl-\hat{Al}-Cl$ angle of $\sim 112°$ (**29**). Work by the groups of Beattie and Ogden at Southampton University, however, indicated the planar structure **30**. Electron diffraction data on the gas phase molecule suggest a planar or near-planar geometry, in line with the matrix results of Beattie and Ogden. The discrepancy was finally resolved when it was realized that the non-planarity of $AlCl_3$ in some matrix studies actually results from weak complex formation [41].

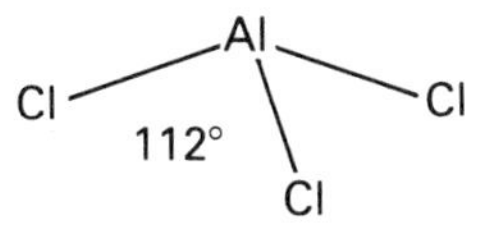

(**29**) 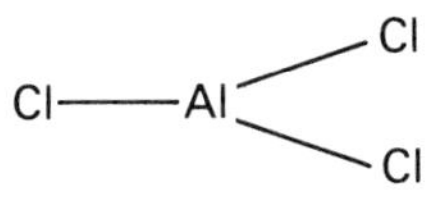                                             (**30**)

A recent infrared study on matrix-isolated alkali metal hexafluorouranates(V), $MUF_6$ (M = K, Rb or Cs), has emphasized how the nature of the matrix material may influence the structure of the molecules isolated within it [42]. Fig. 7.8 shows the infrared spectrum of $CsUF_6$ isolated in both an argon and a nitrogen matrix. The difference between the spectra can be explained if the molecule adopts a $C_{3v}$ molecular symmetry in argon and a $C_{2v}$ molecular symmetry in nitrogen. This change in symmetry implies a tridentate interaction in argon where the $Cs^+$ cation is co-ordinated to a face of the octahedral $UF_6^-$ anion (**31**) as opposed to a bidentate interaction in nitrogen, where the $Cs^+$ cation is co-ordinated to an edge of the $UF_6^-$ octahedron (**32**).

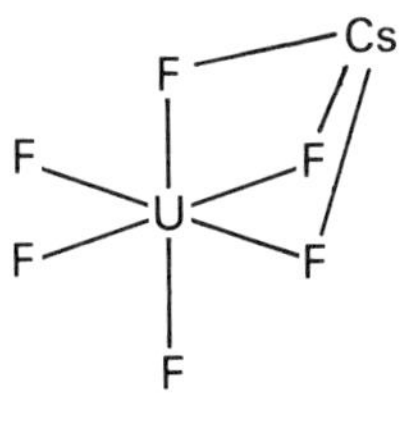

(**31**) 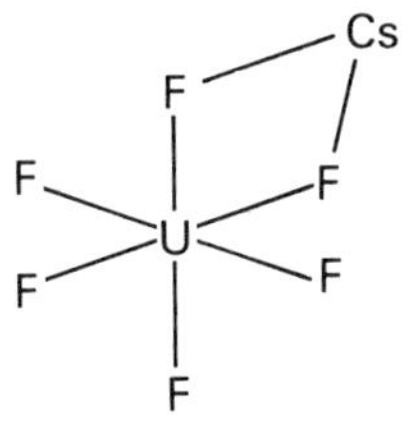                                             (**32**)

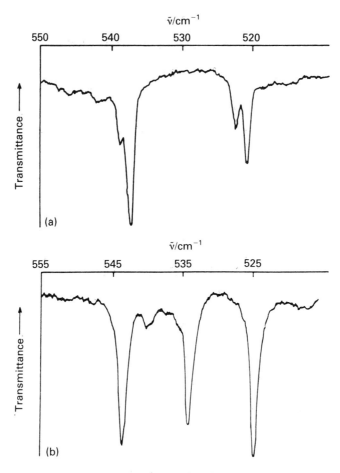

Fig. 7.8 — Infrared spectra of CsUF$_6$, (a) in an argon matrix, (b) in a nitrogen matrix (both at 14 K). Reproduced with permission from Arthers *et al.*, *J. Chem. Soc., Dalton Trans.*, 711 [©] 1984 Royal Society of Chemistry.

## 7.7   REACTIONS OF METAL PORPHYRINS

Reactions between metal porphyrins and dioxygen provide extreme examples of complex formation. For some years it has been known that base-complexed metal porphyrins — for example, the pyridine complex of (tetraphenylporphyrinato) manganese(II) — will undergo reversible oxygenation in solution [43]:

$$Mn(TPP)Py + O_2 \rightleftharpoons Mn(TPP)O_2 + Py$$

The reaction has been monitored by ESR and electronic spectroscopy, and these measurements suggest that the O$_2$ ligand is bound in a side-on manner to the manganese atom (**33**).

**(33)**

However, it is only by matrix isolation that the reactions between base-free metal porphyrins and dioxygen have been discovered, most of this work being carried out by the group of Nakamoto at Wisconsin. These workers have produced base-free metal porphyrins in high-temperature vapours by heating the base-complexed species in a furnace until complete dissociation has occurred, and all the base molecules have been lost. By increasing the furnace temperature — typically to 400–550 K — the metal porphyrin complex may itself be vaporized, and this vapour co-condensed with $O_2$/argon mixtures. Thus the side-on co-ordination of $O_2$ to manganese porphyrins has been confirmed by infrared spectroscopy [44]. By contrast, $O_2$ binds in bent end-on manner to cobalt porphyrins [45] (**34**).

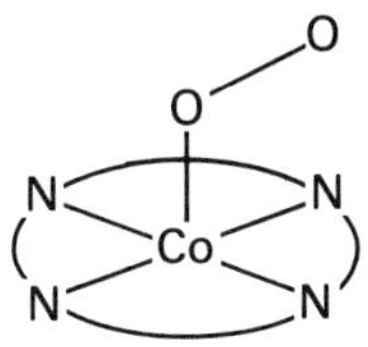

(**34**)

The reactions of iron porphyrin complexes with $O_2$ are somewhat more complicated. Here it appears that two isomers of the complexes Fe(porph) ($O_2$) (where porph = tetraphenylporphyrinato or octaethylporphyrinato) are generated, one with side-on (**35**) and the other with end-on (**36**) bonding of $O_2$ [46]. These isomers may be interconverted thermally, whereas laser photolysis at 406.7 nm causes cleavage of the O–O bond and formation of a complex OFe(porph) which contains the ferryl O=Fe group [47]. This product has been characterized by means of its resonance Raman spectrum.

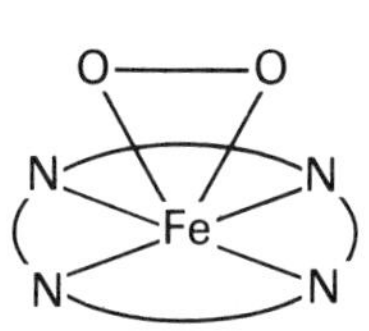

(**35**)

$\nu_{o-o} \sim 1105$ cm$^{-1}$

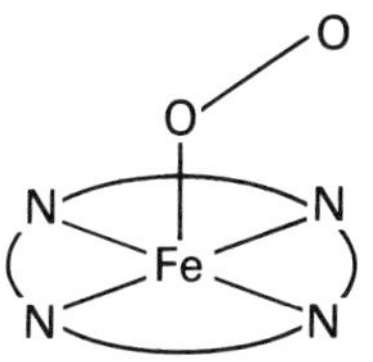

(**36**)

$\nu_{o-o} \sim 1190$ cm$^{-1}$

## 7.8  SUMMARY

This chapter describes some of the areas of chemistry in which progress has been made in characterizing the species present in high-temperature vapours, and in studying their chemical reactivity. Such studies cover an enormous range of molecules from the simple diatomics such as silicon monoxide and the aluminium monohalides through to the complicated metal porphyrin systems.

Matrix isolation has proved to be a good way of trapping high-temperature vapour species, and is often ideal for studying their reactions. For example, several reactions of SiO have been observed in this way. However, the results of matrix-isolation experiments are not always unambiguous. This point is illustrated by the structural studies on the transition metal dihalides, on $AlCl_3$ and on the metal hexafluorouranate(V) molecules. Here it appears that the matrix-isolated species is structurally affected by the matrix host material, so it is difficult to say exactly which structure corresponds to that of the gas phase molecule.

Thus gas phase measurements — including electron diffraction and mass spectrometry — assume an importance. However, these experiments, too, are not always totally clear. Electron diffraction data, on high-temperature molecules, are often difficult to interpret, because of the high vibrational excitation of the molecule, leading to the so-called shrinkage effect. Thus, for example, excitation of the bending mode of a triatomic molecule makes it difficult to say whether the *equilibrium* geometry of the molecule is linear or slightly bent. The increasingly sophisticated analysis of electron diffraction results will help to overcome this problem. However, a combination of gas phase and matrix-isolation measurements is the best way to characterize a molecule present in a high-temperature vapour. In particular for simple molecules, it is desirable that these experimental measurements should be backed up by reliable theoretical *ab initio* calculations. As more elaborate theoretical calculations become possible, this approach will be extended to a wider range of molecules.

It is likely that the use of high-temperature molecules in preparative chemistry — exemplified in this chapter by the reactions of AlCl — will become more widespread. Some products of these reactions may find use in other areas of preparative chemistry. It is possible, for example, that new organometallic molecules will be made in this way, which will open up new synthetic possibilities.

## REFERENCES

[1] J. S. Anderson and J. S. Ogden, *J. Chem. Phys.* (1969), **51**, 4189.
[2] R. F. Porter, W. A. Chupka and M. G. Inghram, *J. Chem. Phys.* (1955), **23**, 216.
[3] R. M. Atkins and P.L. Timms, *Spectrochim. Acta part A.* (1977), **33A**, 853.
[4] H. Schnöckel, *Angew. Chem. Int. Ed. Engl.* (1978), **17**, 616.
[5] H. Schnöckel, *Z. Anorg. Allg. Chem.* (1980), **460**, 37.
[6] H. Schnöckel, *Angew. Chem. Int. Ed. Engl.* (1980), **19**, 323.
[7] T. Mehner, H. Schnöckel, M. J. Almond and A. J. Downs, *J. Chem. Soc., Chem. Commun.* (1988), 118.
[8] F. C. Wyse and W. Gordy, *J. Chem. Phys.* (1972), **56**, 2130.
[9] H. Schnöckel, *Z. Naturforsch.B.* (1976), **31B**, 1291.
[10] R. Ahlrichs, L. Zhengyan and H. Schnöckel, *Z. Anorg. Allg. Chem.* (1984), **519**, 155.
[11] H. Schnöckel, *J. Mol. Struct.* (1978), **50**, 267.
[12] L. Zhengyan, H. Janssen, R. Mattes, H. Schnöckel and B. Krebs, *Z. Anorg. Allg. Chem.* (1984), **513**, 67.

[13] H. Schnöckel, M. Leimkühler, R. Lotz and R. Mattes, *Angew. Chem. Int. Ed. Engl.* (1986), **25**, 921.

[14] B. Bak, N. A. Kristiansen and H. Svanholt, *Acta Chem. Scand. A.* (1982), **36A**, 1.

[15] M. Binnewies, M. Lakenbrink and H. Schnöckel, *Z. Anorg. Allg. Chem.* (1983), **497**, 7; H. Schnöckel and S. Schunck *Z. Anorg. Allg. Chem.* (1987), **548**, 161.

[16] H. Schnöckel and M. Lakenbrink, *Z. Anorg. Allg. Chem.* (1983), **507**, 70.

[17] R. Ahlrichs, R. Becherer, M. Binnewies, H. Borrmann, M. Lakenbrink, S. Schunk and H. Schnöckel, *J. Amer. Chem. Soc.* (1986), **108**, 7905.

[18] R. Ahlrichs, C. Ehrhardt, M. Lakenbrink, S. Schunck and H. Schnöckel, *J. Amer. Chem. Soc.* (1986), **108**, 3596.

[19] V. A. Kulikov, V. V. Ugarov and N. G. Rambidi, *Zh. Strukt. Khim.* (1981), **22**, 166.

[20] N. M. Roddatis, S. M. Tolmachev, V. V. Ugarov, Y. S. Ezhov and N. G. Rambidi, *Zh. Strukt. Khim.* (1974), **15**, 693.

[21] S. A. Arthers, I. R. Beattie, R. A. Gomme, P. J. Jones and J. S. Ogden, *J. Chem. Soc., Dalton Trans.* (1983), **1461**; I. R. Beattie and J. E. Parkinson, *J. Chem. Soc., Dalton Trans.* (1984), 1363.

[22] L. Bencivenni, H. M. Nagarathna, K. A. Gingerich and R. Teghil, *J. Chem. Phys.* (1984), **81**, 3415.

[23] V. P. Spiridonov, A. N. Khodchenkov and P. A. Akishin, *Vestn. Mosk. Univ.* (1965), **6**, 34.

[24] J. S. Ogden and S. J. Williams, *J. Chem. Soc., Dalton Trans.* (1982), 825.

[25] L. Bencivenni and K. A. Gingerich, *J. Mol. Struct.* (1983), **99**, 23; L. Bencivenni and K. A. Ginerich, *J. Mol. Struct.* (1983), **98**, 195.

[26] J. R. Beattie and J. E. Parkinson, *J. Chem. Soc., Dalton Trans.* (1983), 1185.

[27] G. Ritzhaupt, K. Consani and J. P. Devlin, *J. Chem. Phys.* (1985), **82**, 1167.

[28] S. J. David and B. S. Ault, *Inorg. Chem.* (1984), **23**, 1211; *Inorg. Chem.* (1985), **24**, 1048.

[29] W. D. Hewett, Jr., J. H. Newton and W. Weltner, Jr., *J. Phys. Chem.* (1975), **79**, 2640.

[30] D. W. Green and K. M. Ervin, *J. Mol. Spectrosc.* (1981), **89**, 145.

[31] I. Hargittai, M. Hargittai, V. P. Spiridonov and E. V. Erokhin, *J. Mol. Struct.* (1971), **8**, 31; A., A. Ivanov, A. V. Demidov, N. I. Popenko, E. Z. Zasorin, V. P. Spiridonov and I. Hargittai, *J. Mol. Struct.* (1980), **63**, 121.

[32] R. J. Ackermann and E. G. Rauh, *J. Phys. Chem.* (1963), **67**, 2596.

[33] R. J. Mawhorter, M. Fink and J. G. Hartley, J. Chem. Phys. (1985), **83**, 4418.

[34] Z. K. Ismail, R. H. Hauge and J. L. Hargrave, *J. Mol. Spectrosc.* (1975), **54**, 402.

[35] G. E. Leroi, T. C. James, J. T. Hougen and W. Klemperer, *J. Chem. Phys.* (1962), **36**, 2879.

[36] R. A. Frey, R. D. Werden and H. H. Günthard, *J. Mol. Spectrosc.* (1970), **35**, 260; J. W. Hastie, R. H. Hauge and J. L. Margrave, *High Temp. Sci.* (1971), **3**, 257; K. R. Thompson and K. D. Carlson, *J. Chem. Phys.* (1968), **49**, 4379.

[37] A. Büchler, J. L. Stauffer and W. Klemperer, *J. Amer. Chem. Soc.* (1964), **86**, 4544.

[38] I. Hargittai, J. Tremmel and G. Schultz, *J. Mol. Struct.* (1975), **26**, 116; E. Vajda, J. Tremmel and I. Hargittai, *J. Mol. Struct.* (1978), **44**, 101.

[39] D. W. Green, D. P. McDermott and A. Bergman, *J. Mol. Spectrosc.* (1983), **98**, 111.

[40] M. Hargittai and I. Hargittai, *J. Mol. Spectrosc.* (1984), **108**, 155.

[41] I. R. Beattie, H. E. Blayden and J. S. Ogden, *J. Chem. Phys.* (1976), **64**, 909; J. S. Shirk and A. E. Shirk, *J. Chem. Phys.* (1976), **64**, 910.

[42] S. A. Arthers, I. R. Beattie and P. J. Jones, *J. Chem. Soc., Dalton Trans.* (1984), 711.

[43] B. M. Hoffman, C. J. Weschler and F. Basolo, *J. Amer. Chem. Soc.* (1976), **98**, 5473.

[44] M. W. Urban, K. Nakamoto and F. Basolo, *Inorg. Chem.* (1982), **21**, 3406; K. Nakamoto, T. Watanabe, T. Ama and M. W. Urban, *J. Amer. Chem. Soc.* (1982), **104**, 3744.

[45] M. Kozuka and K. Nakamoto, *J. Amer. Chem. Soc.* (1981), **103**, 2162.

[46] K. Nakamoto, T. Watanabe, T. Ama and M. W. Urban, *J. Amer. Chem. Soc.* (1982), **104**, 3744; T. Watanabe, T. Ama and K. Nakamoto, *J. Phys. Chem.* (1984), **88**, 441.

[47] K. Bajdor and K. Nakamoto, *J. Amer. Chem. Soc.* (1984), **106**, 3045.

# 8

# Ions and radicals

## 8.1  INTRODUCTION

Molecular ions and radicals show, in general, a high chemical reactivity, and most exist only transiently under normal conditions. Because of their reactivity, such species have been studied principally in flow systems, and, in recent years, matrix-isolation experiments have also assumed an increasing importance. Several routes to radical formation exist; photolysis, pyrolysis, or the action of a discharge source — usually either microwave or radio frequency — have all been employed. Bimolecular reactions leading to homolytic bond fission will also yield radicals.

Various methods are known by which molecular ions may be generated. Obviously any reaction resulting in heterolytic bond fission will yield ions, though such reactions are encountered normally only in solutions in polar solvents. Molecular cations may also be formed by ionization, where electrons are removed from a parent molecule. Ionization may be brought about by photolysis, radiolysis (e.g. with $\gamma$-rays) or electron bombardment. Similar processes, resulting in electron capture, yield molecular anions.

The use of microwave or radio frequency discharges for depositing thin layers of material, or for preparing unusual molecules, is well known. Examples are the deposition of amorphous hydrogenated silicon layers from silane discharges, and the formation of polyboron halide clusters from boron trihalide precursors. Extremely complex arrays of species, including ions and neutral radicals, are produced in the plasmas generated by discharges, and recently attempts have been made to identify some of these plasma species.

## 8.2  FORMATION OF RADICALS

Radicals are produced by homolytic bond fission, which is the preferred form of bond-breaking for all gas phase molecules. In principle, given sufficient energy, any covalent bond will dissociate to produce radicals on thermolysis. However, some bonds — for example the O—O bonds of organic peroxides — undergo very facile homolysis. Thus methyl radicals are formed on thermal decomposition of diacetyl peroxide or di-$t$-butyl peroxide.

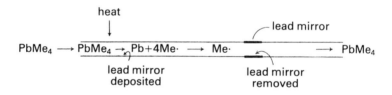

Several organometallic molecules also readily decompose to yield organic radicals. In 1926, Paneth [1] first demonstrated the presence of methyl radicals in the gas phase, following pyrolysis of tetramethyl lead vapour in a glass tube. He found that methyl radicals, produced in this way, react with a lead mirror further down the tube, removing the lead mirror, and regenerating Pb(CH₃)₄ (see Fig. 8.1).

Fig. 8.1 — Schematic diagram of the Paneth lead mirror experiment.

Methyl radicals may also be formed by photolysis of acetone vapour at wavelengths close to 310 nm. The ESR spectrum of gas phase methyl radicals, generated by either pyrolytic or thermolytic means, has been recorded by flowing the radical-containing gas stream through an ESR spectrometer cavity. The principal feature of this spectrum is a quartet of lines with the intensity ratio 1:3:3:1, indicating coupling of the odd electron to three equivalent hydrogen atoms [2].

Chemical reactions may also lead to bond homolysis. Thus matrix-isolated methyl radicals are formed (as described in section 4.7) when lithium atoms are co-condensed with $CH_3Br$ and excess argon [3].

Another route which has been used to generate both gas phase and matrix-isolated radicals is abstraction of hydrogen atoms from either inorganic or organic hydrides on reaction with fluorine atoms. Reaction of F atoms — generated by passing a stream of $CF_4/Ar$ or $NF_3/Ar$ through a microwave discharge cavity — with methane in the gas phase generates vibrationally excited HF. The other principal product is likely to be $CH_3$, but direct detection of this radical, in the gas phase, under these conditions, has proved to be difficult. Secondary fluorine atom reactions lead to other products, including the CH radical whose far infrared spectrum has been recorded directly by the technique of laser magnetic resonance [4]. If, however, the products of the reaction of $CH_4$ with F atoms are frozen with excess Ar at 14 K, then the infrared spectrum of the resulting deposit shows clearly the presence of matrix-isolated HF and $CH_3$, alongside a $CH_3$ . . . HF complex [5]. Thus it appears that a substantial $CH_3$ concentration *is* formed in the gas phase by this reaction. Methyl radicals may also be formed by the interaction of $CH_4$ and electronically excited Ar atoms, produced in a microwave discharge [6].

A more complicated system is that involving methanol, $CH_3OH$, and flourine atoms. In principle, two different radicals — $CH_2OH$ or $CH_3O$ — may be envisaged as primary products from this reaction. Both of these radicals play a significant role in hydrocarbon combustion, in air pollution and are also likely to be involved in the chemistry of interstellar matter. Studies of their chemistry, therefore, assume an importance in a variety of contexts. It appears that $CH_3O$ is the most abundant primary product formed by high-energy, e.g. X-ray, irradiation of solid $CH_3OD$ at 4.2 K [7]. The ESR spectrum of such a sample clearly demonstrates the presence of $CH_3O$, alongside smaller yields of $CH_2OD$ and $CH_3$. On warming to 77 K, the concentration of $CH_3O$ decreases and that of $CH_2OD$ increases, presumably as a result of a hydrogen abstraction reaction.

$$CH_3OD \xrightarrow[\text{4.2 K}]{\text{X-ray}} CH_3O + D$$

$$CH_3O + CH_3OD \xrightarrow{\text{77 K}} CH_3OH + CH_2OD$$

By contrast, vacuum ultraviolet photolysis (at 147 nm) of $CH_3OH$ isolated in argon or nitrogen matrices at 14 K yields, on the evidence of the infrared spectrum, $CH_2OH$ as the principal product [8].

$$CH_3OH \xrightarrow[\text{Ar matrix, 14 K}]{\lambda = 147\,\text{nm}} CH_2OH + H$$

The reaction of F atoms with $CH_3OH$ generates in the gas phase, $CH_3O$, the rotational spectrum of which has been measured by laser magnetic resonance [9]. This study provides the first certain spectroscopic detection of the methoxy radical. On the other hand, when the products of the interaction between methanol and either excited Ar atoms or F atoms are frozen in a large excess of Ar at 14 K, prominent infrared absorptions of $CH_2OH$ appear. There is some evidence for the stabilization of $CH_3O$, but the reaction channel producing $CH_2OH$ predominates [10]. Perhaps the selectivity of the laser magnetic resonance technique has allowed the $CH_2OH$ radical to escape detection in the gas phase. Moreover, it is known that $CH_2OH$ is thermodynamically more stable than $CH_3O$, by some 20–40 kJ mol$^{-1}$, so it is possible that the process of condensing the reaction products in the matrix increases the yield of the more stable isomer.

### 8.3  FORMATION OF RADICAL CATIONS

The cations of most neutral molecules are radicals, and are hence amenable to ESR study. A successful experimental approach has been to trap such ions in rare gas (particularly neon) matrices, where ESR detection is made. The extreme chemical inertness of neon makes it an ideal host for trapping the more reactive radical cations, with high electron affinities. Furthermore, ESR studies of the $^{13}CO^+$ and $H_2O^+$ ions, both in neon matrices and in the gas phase, indicate that the neon matrix shifts of the nuclear hyperfine parameters are in the order of only 2–3%. Less reactive cations, especially larger organic radical cations, have, however, been studied successfully by ESR spectroscopy in freon or hydrocarbon glasses.

Generation of radical cations is normally achieved by vacuum ultraviolet photolysis, $\gamma$-radiolysis or electron bombardment of the neutral parent molecule either following, or immediately prior to, condensation with excess inert gas. Thus the isoelectronic ions $N_2^+$ and $CO^+$ have been generated by bombardment of neon matrices [11]. Fig. 8.2 shows the ESR spectrum of matrix-isolated $^{14}N_2^+$. The quintet of lines results from the interaction of the unpaired electron with the two equivalent $I = 1$ $^{14}N$ nuclei.

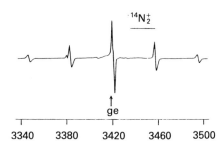

Fig. 8.2 — ESR spectrum of $^{14}N_2^+$ in a neon matrix at 4 K. Reproduced with permission from Knight, *Acc. Chem. Res.*, **19**, 313 ©1986 American Chemical Society.

A question which must be answered about such processes is, what happens to the electrons? This problem has been explained at least partially by experiments in which cations such as $N_2^+$ are generated in neon matrices doped with $F_2$ [12]. In these experiments the ESR spectrum of the radical anion $F_2^-$ is also observed. It appears that both the radical anion and the radical cation are essentially 'free' and they do not exist as ion pairs in the matrix, since no $^{19}F$ hyperfine splitting is observed in the ESR spectrum of the cation. It is likely that in all such ionization experiments the electrons are held in anion traps within the matrix.

Low-temperature glasses have been used to stabilize a variety of inorganic and organic radical cations, which have been studied by ESR spectroscopy. Thus $\gamma$-radiolysis of stannane, $SnH_4$, trapped in a freon glass, gives rise to two isomers of the $SnH_4^+$ cation, having $C_{2v}$ (1) and $C_{3v}$ (2) symmetry [13]. These structures arise from a Jahn–Teller distortion, from tetrahedral symmetry, caused by the presence of an unpaired electron. It is noteworthy that two distorted structures are seen in almost equal abundance, although the form with $C_{3v}$ symmetry is somewhat more stable. By contrast, only one form of the anion $SnH_4^-$ is produced from $SnH_4$ by electron capture in a neopentane matrix, and this shows $C_{2v}$ symmetry [14].

(1)                              (2)

$\gamma$-Radiolysis of $N_2O_4$ trapped in a $CFCl_3$ glass at 77 K gives the radical cation $N_2O_4^+$ is produced from $SnH_4$ [15].Interestingly, the ESR spectrum of this species closely resembles that of the radical $NO_2^{\bullet}$. Thus it is proposed that $N_2O_4^+$ consists of a bent $NO_2^{\bullet}$ radical strongly perturbed by a linear $NO_2^+$ cation (3). This conclusion is supported by the results of theoretical *ab initio* calculations.

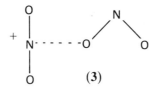

(3)

Various *n*-alkane radical cations, of general formula $[H(CH_2)_nH]^+$, have been stabilized in $SF_6$ matrices [16]. These cations show $\sigma$-delocalization of the unpaired electron, much of the spin density residing on the 1s orbitals of the terminal hydrogen atoms (4). The ion $[C_2H_6]^+$ and the cycloalkane cations $[C_3H_6]^+$, $[C_4H_8]^+$ and $[C_5H_{10}]^+$ all show Jahn–Teller distortions. Although the ESR spectrum of $[CH_4^+]$ trapped in a neon matrix at 4 K shows four equivalent hydrogen atoms, this equivalence is almost certainly due to rapid averaging. Experiments on $[CH_2D_2]^+$ show that the static structure of $CH_4^+$ has $C_{2v}$ symmetry, in common with one form of $SnH_4^+$ (1) [17].

(4)

More complex hydrocarbon cations readily undergo rearrangement. For example, the bicyclo-[2.1.0]-pentane cation (**5**) isomerizes in a freon glass, even at 4 K, to the cyclopentene cation (**6**) [18]. ESR measurements offer an ideal means of monitoring such isomerizations.

(5)                                                   (6)

Knight and his co-workers at Furman University have extended the scope of their studies to include matrix isolation of the radical cations derived from high-temperature molecules. Part of their strategy has centred on a technique known as reactive laser sputtering, in which a reagent gas is passed over a metal target which is undergoing sputtering under the action of focused laser pulses. In some cases this process is, in itself, sufficient to produce molecular ions — $AlH^+$, for example, having been generated in this way [19]. For most species, however, reactive laser sputtering generates neutral parent molecules in the gas phase, which are then ionized by photolysis or electron bombardment, this treatment being required to generate the cation $AlF^+$ [20]. An alternative process is to ionize vapour phase neutral molecules, which have been generated by vaporization of solid samples. Thus the cation $SiO^+$ which is isoelectronic with $AlF^+$ has been produced by ionization of SiO, prepared, in turn, by vaporization of silicon oxide samples at about 2000 K [21]. Characterization of $SiO^+$ and $AlF^+$ by ESR spectroscopy following trapping in neon matrices has confirmed that both are $^2\Sigma$ radicals.

## 8.4  HALOBENZENE RADICAL CATIONS

One group of molecular ions which has received much attention is the radical cations derived from halobenzene molecules. Two routes have been used to generate such ions. These are ultraviolet or vacuum ultraviolet irradiation of matrix-isolated halobenzene molecules, or co-condensation of halobenzene molecules with a stream of argon gas, which has been excited by passage through a discharge tube. The latter method is more efficient, as can be seen from Fig. 8.3, which contrasts the yield of the cation $[para\text{-}C_6H_4Cl_2]^+$ obtained by the two different approaches [22].

An aim of this work, which has been carried out by the groups of Andrews at University of Virginia and of Bondybey at the Bell Telephone Laboratories, has been to study the structures of the ground-state and low-lying electronically excited states of the radical cations. Information on the ground state has come principally from electronic emission spectra, while electronic absorption spectra yield information on the excited electronic states. The emission spectra of $C_6F_6^+$ (**7**) and $[1,3,5\text{-}C_6H_3F_3]^+$ (**8**) show that the vibrational structure of the ground electronic states of these cations is irregular and complicated, a finding consistent with a

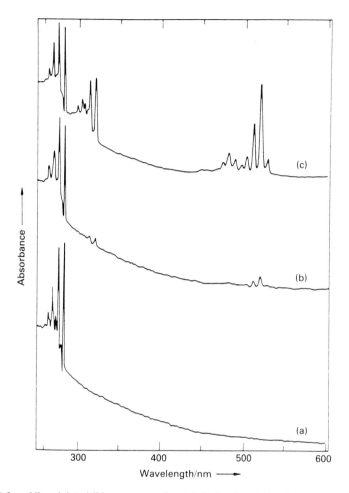

Fig. 8.3 — Ultraviolet-visible spectrum of matrix-isolated *p*-dichlorobenzene and its radical cation: (a) before photolysis; (b) after 15 min photolysis at λ=220–1000 nm; (c) after Ar-resonance photolysis. Reproduced with permission from Friedman *et al.*, *J. Phys. Chem.*, **88**, 1944 ©1984 American Chemical Society.

Jahn–Teller distortion of the ground state [23]. By contrast, the absorption spectra of **8** show that the excited state is characterized by an extremely sharp and regular vibrational structure. The observation of a Jahn–Teller distortion for these cations is expected, since the high symmetry would allow a degenerate ground state, and the cations possess an unpaired electron, so distortion occurs to lift the degeneracy.

(7)                                                    (8)

For less symmetric halobenzene cations, such as $[m\text{-}C_6H_4F_2]^+$ (**9**), the situation is somewhat different [24]. The reduced symmetry gives rise to a non-degenerate ground state. Thus there can be no Jahn–Teller distortion, and the emission spectra point to a ground state where the vibrational structure is mostly regular. However, the absorption spectra show that the vibrational structure for the upper electronic state is very irregular. The most plausible explanation is that two nearly degenerate excited states exist, and that these perturb each other.

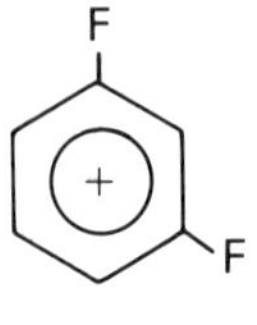

(**9**)

## 8.5  ION-NEUTRAL REACTIONS

One of the most interesting and useful applications of the matrix-isolation approach to investigating molecular ions is the study of ion-neutral rections. Such reactions are very difficult to study by other experimental procedures.

If the ion $CO^+$ — formed by either electron bombardment or photolysis — is allowed to react with neutral CO during co-condensation with excess neon at 4 K, then the product is the matrix-isolated ion $C_2O_2^+$ [25]. It appears that the ion-neutral reaction

$$CO + CO^+ \xrightarrow[\text{4 K}]{\text{Ne}} C_2O_2^+$$

takes place during condensation of the matrix. The structure of the $C_2O_2^+$ product has been determined from experimental ESR spectroscopic measurements, backed up by theoretical *ab initio* calculations. It is found to be a planar *trans* molecule [**10**] with a $C-\hat{C}-O$ bond angle of 141° and C–C and C–O bond lengths of 1.58 and 1.41 Å respectively. The unpaired electron is delocalized over the entire molecular framework.

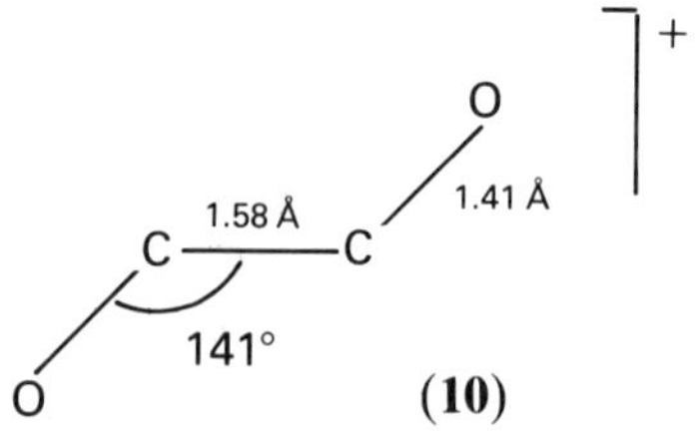

(**10**)

A similar product, $N_2CO^+$, is formed by the reaction of $CO^+$ with $N_2$. Fig. 8.4 shows the ESR spectrum of the isotopomer $^{15}N_2CO^+$. The presence of a doublet of

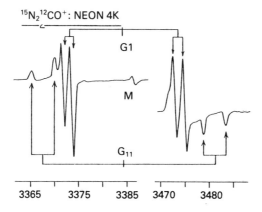

Fig. 8.4 — ESR spectrum of $^{15}N_2CO^+$ in a neon matrix at 4 K. Reproduced with permission from Knight, *Acc. Chem. Res.*, **19**, 313 ©1986 American Chemical Society.

doublets can clearly be discerned, and this observation points to the interaction of the unpaired electron with two inequivalent $I = \frac{1}{2}$ $^{15}N$ nuclei. More recently the ion $N_4^+$ has been prepared by the reaction

$$N_2^+ + N_2 \xrightarrow[4\,K]{Ne} N_4^+$$

during co-condensation with excess neon at 4 K [26]. Unlike its isoelectronic analogue $C_2O_2^+$, $N_4^+$ is a linear species (**11**).

$$[N-N-N-N]^+$$

$$(11)$$

It has a $^2\Sigma$ ground state, and the unpaired electron resides principally on the two inner nitrogen atoms. Again this structure has been determined by a combination of ESR measurements and *ab initio* calculations.

### 8.6 PHOTOCHEMISTRY OF MOLECULAR IONS

Low-energy visible light photolysis of a matrix containing both radical cations and neutral radicals has been found to eliminate the ESR signals of the trapped ions, but does not affect the intensity of the ESR signals of most neutral radicals. This photobleaching effect allows the two types of species to be easily distinguished. Thus, for example, the ESR signals resulting from the matrix-isolated radicals $NH_3^+$ and $NH_2$ — produced by electron bombardment of $NH_3$ — have been assigned [27], since the $NH_3^+$ signals markedly decrease on visible irradiation.

The mechanism of this process appears to rely on the fact that while visible photons are not energetic enough to break most chemical bonds, they can ionize the anion traps (i.e. the sites where electrons are held) within the matrix. Thus electrons are released. These can move freely within the solid matrix lattice, and will neutralize all types of radical cation in a non-selective manner. The process is illustrated schematically in Fig. 8.5.

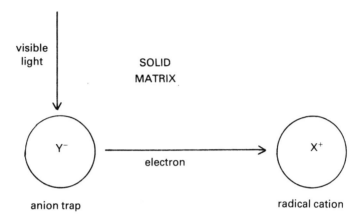

Fig. 8.5 — Neutralization of a radical cation by electrons released from an anion trap by visible photolysis.

Other radical cations — generally those with more complicated structures — undergo rearrangement reactions on visible photolysis. Thus the *o*- and *m*-dichloro-benzene cations (**12, 13**) rearrange to the *p*-dichlorobenzene cation (**14**) on visible photolysis [28]. It is proposed that this rearrangement occurs via a bridged chlorinium ion (**15**), which allows migration of the chlorine atom around the ring.

Photo-induced rearrangement reactions are not limited to halogenated organic radical cations. Visible photolysis of the cycloheptatriene cation (**16**) trapped in an argon matrix causes conversion to the toluene cation (**17**). As illustrated in Fig. 8.6 the reaction has been monitored by observing the visible absorption spectrum of the

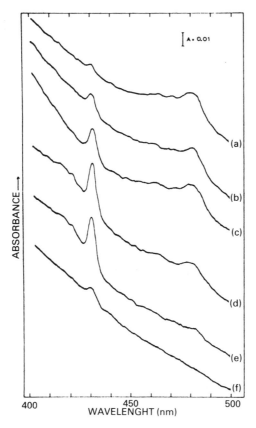

Fig. 8.6 — Visible spectrum of cycloheptatriene in an argon matrix: (a) after deposition (b) after photolysis at λ>590 nm; (c) after photolysis at λ>520 nm; (d) after photolysis at λ>470 nm; (e) after photolysis at λ>420 nm; (f) after photolysis at λ>290 nm (all photolyses carried out for 15 min). Reproduced with permission from Andrews and Keelan, *J. Am. Chem. Soc.*, **102**, 5732 ©1980 American Chemical Society.

matrix, where the cycloheptatriene cation (**16**) shows an absorption at 480 nm and the toluene cation (**17**) an absorption at 430 nm [29a]. Further photolytic rearrangements of **16** and **17** are possible. Visible photolysis of either cation will also yield the methylene cyclohexadiene cation (**18**). It appears that these rearrangements occur via a series of photo-equilibria probably involving the norcaradiene cation (**19**). In part, the interest in these matrix studies on $C_7H_8^+$ ions stems from the light shed on the well-known McLafferty rearrangement of such ions in the gas phase during mass spectrometric detection. As a result of mass spectrometric studies of partially deuterated toluene molecules, McLafferty and his co-workers have proposed that **18** and **19** are intermediates in the gas phase rearrangement of the toluene cation (**17**) [30]. The direct observation of rearrangement reactions of $C_7H_8^+$ in solid matrices adds considerable support to McLafferty's proposed mechanisms.

(**16**)                    (**19**)                    (**18**)                    (**17**)

In these studies on photochemical rearrangements of matrix-isolated cations, the matrix is doped with dichloromethane, $CH_2Cl_2$, which acts as an anion trap. This means of trapping the released electrons not only increases the yield of cations formed, but also serves to inhibit the neutralization reactions mentioned earlier in this section, since the $CH_2Cl_2$ molecule holds the electrons too strongly for them to be readily released on visible photolysis.

## 8.7  PLASMA DISCHARGE SYSTEMS

Plasmas, produced by discharges, contain both ionic and neutral short-lived fragments. The identification of these fragments is of great importance in the research on the growth and properties of thin films. For example, amorphous hydrogenated silicon (a-Si:H), a potentially important solar cell material, is deposited from silane discharges, in which neutral radicals are known to be the dominant species. Charcterization of these radicals has proved, however, to be extremely difficult.

Silane plasmas have been studied directly by gas phase infrared and ultraviolet absorption and emission spectroscopy [31] and by mass spectrometry [32]. The results of these experiments have shown the chemical complexity of silane plasmas. While the IR and UV spectra demonstrate clearly the presence of the neutral radical SiH, alongside the ion $SiH^+$ [31], the dominant species observed by mass spectrometry are the ions $SiH_2^+$ and $SiH_3^+$, although the radicals SiH, $SiH_2$ and $SiH_3$ have also been detected [32]. It is still not clear whether silylene, $SiH_2$, or the silyl radical, $SiH_3$, is the principal species involved in a-Si:H deposition from silane plasmas.

Recently an attempt has been made to identify the radicals present in a silane discharge by trapping them in an argon matrix, then using infrared spectroscopic detection [33]. Although the results of this preliminary study do not provide a wholly unambiguous assignment of the radicals present in the plasma, it is clear that both $SiH_3$ and $SiH_2$ are produced. Furthermore, these preliminary results show the efficiency of matrix isolation as a method of characterizing free radicals produced in a plasma discharge. It is hoped that more refined experiments will provide a definitive assignment of the radicals present, and that it will prove possible to make quantitative measurements.

The use of discharge systems in preparative chemistry is exemplified by the synthesis of boron subhalides from boron trihalide precursors. In 1925, Alfred Stock first reported the isolation of one drop of diboron tetrachloride, $Cl_2BBCl_2$, (**20**) from the zinc discharge reduction of liquid boron trichloride [34a]. Some years later it was demonstrated by Schlesinger and his co-workers that a much improved yield of $B_2Cl_4$ is obtained if $BCl_3$ vapour is passed at low pressure through a mercury discharge [34b]. A schematic representation of the apparatus is given in Fig. 8.7. It appears that BCl is the intermediate in this reaction, and that $B_2Cl_4$ is produced by insertion of

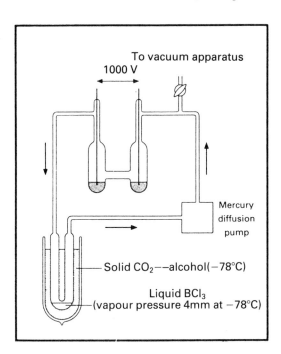

Fig. 8.7 — Diagram of the apparatus used to prepare $B_2Cl_4$ from $BCl_3$. Reproduced with permission from Massey, *Chem. Br.*, **16**, 588 ©1980 Roayl Society of Chemistry.

BCl into a B−Cl bond of $BCl_3$ [35] (compare with the insertion reactions of the molecule AlCl mentioned in section 7.3). Likewise $B_2Br_4$ may be formed from $BBr_3$.

$$2Hg^* + BCl_3 \xrightarrow[\text{discharge}]{\text{plasma}} Hg_2Cl_2 + [BCl]$$

$$[BCl] + BCl_3 \xrightarrow{\text{insertion}} B_2Cl_4$$

These molecules are of interest because they contain an unusual direct boron–boron bond. $B_2Cl_4$ adopts a staggered $D_{2d}$ structure (**20a**) in the gas phase and a planar $D_{2h}$ structure (**20b**) in the crystalline state.

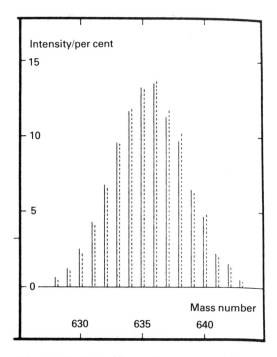

(**20a**)                    (**20b**)

Two important by-products formed in the mercury discharge synthesis of $B_2Cl_4$ are the pale-yellow tetrahedral cluster $B_4Cl_4$ (**21**) and the yellow–orange $B_9Cl_9$. Likewise, passing $B_2Br_4$ vapour through a silent electric discharge yields the red cluster $B_9Br_9$. Other boron subhalide clusters have been prepared by thermal decomposition of $B_2Cl_4$ and $B_2Br_4$. On warming from about $-20°C$ to $80°C$, $B_2Cl_4$ forms a range of highly coloured polyboron chlorides $(BCl)_n$ where $n = 8$–$12$; under similar conditions $B_2Br_4$ yields mainly $B_7Br_7$, alongside minor amounts of $B_9Br_9$ and $B_{10}Br_{10}$. It appears that BCl or BBr acts as intermediates in these reactions although the exact mechanisms remain obscure. The intense colours of the polyboron halide

Fig. 8.8 — Observed (——) and calculated (– – – –) mass spectra of the parent molecular ion of $B_7Br_7$. Reproduced with permission from Massey, *Chem. Br.*, **16**, 588 ©1980 Royal Society of Chemistry.

clusters are noteworthy, since it is unusual to find such intense colours outside transition metal chemistry, and the vast majority of boron compounds are colourless.

Mass spectrometry has proved to be an ideal way of characterizing polyboron halide clusters, since each of boron ($^{10}$B, 19.8%; $^{11}$B, 80.2%), chlorine ($^{35}$Cl, 75.5%; $^{37}$Cl, 24.5%) and bromine ($^{79}$Br, 50.5%; $^{81}$Br, 49.5%) have two stable isotopes of high natural abundance. Thus each cluster has a characteristic number and abundance of different isotopomers resulting from the possible combinations of isotopes of the different atoms within the molecule. This is clearly seen in the mass spectrum of the parent molecular ion, since each ion will show a mass spectrum consisting of a particular number of lines, each with a characteristic intensity. Fig. 8.8 shows the mass spectrum of the parent ion $B_7Br_7^+$, derived from the cluster $B_7Br_7$. The solid lines show the observed intensities, and the dashed lines the calculated intensities. An excellent agreement is obtained.

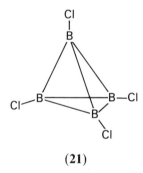

(21)

## 8.8 SUMMARY

This chpater describes some of the studies which have been made on gas phase and matrix-isolated ions and radicals. Matrix isolation has, in recent years, played an important role in such studies. In part this importance has resulted from the development of new routes to matrix-isolated ions and radicals such as the use of windowless resonance lamps for vacuum ultraviolet irradiation, and the ionization techniques developed to produce cations from high-temperature parent molecule precursors. Matrix isolation also allows ion-neutral reactions and photochemical rearrangements of ions to be readily observed. However, as in most other areas, the most reliable results come from a combination of matrix-isolation and gas phase experiments.

Plasma discharge systems are important in the deposition of thin films of materials such as amorphous hydrogenated silicon, and have also been used in preparing novel chemical molecules including the polyboron halide clusters. Such systems are, however, highly complex. Some recent work has focused on attempting to identify the ions and radicals in discharge systems. Both gas phase and matrix-isolation studies have been performed, and it is hoped that this combination of techniques will allow characterization of the species present, and that more quantitative analysis will become possible.

**REFERENCES**

[1] F. A. Paneth and W. Hofeditz, *Ber. Dtsch. Chem. Ges.* (1929), **62**, 1335.

[2] R. W. Fessenden and R. H. Schuler, *J. Chem. Phys.* (1963), **39**, 2147.

[3] L. Andrews and G. C. Pimentel, *J. Chem. Phys.* (1966), **44**, 2527; (1967), **47**, 3637.

[4] J. T. Hougen, J. A. Mucha, D. A. Jennings and K. M. Evenson, *J. Mol. Spectrosc.* (1978), **72**, 463.

[5] M. E. Jacox, *Chem. Phys.* (1979), **42**, 133.

[6] M. E. Jacox, *J. Mol. Spectrosc.* (1977), **66**, 272.

[7] M. Iwasaki and K. Toriyama, *J. Am. Chem. Soc.* (1978), **100**, 1964.

[8] M. E. Jacox and D. E. Milligan, *J. Mol. Spectrosc.* (1973), **47**, 148.

[9] H. E. Radford and D. K. Russell, *J. Chem. Phys.* (1977) **66**, 2222.

[10] M. E. Jacox, *Chem. Phys.* (1981) **59**, 213.

[11] L. B. Knight, Jr. and J. Steadman, *J. Chem. Phys.* (1982), **77**, 1750; L. B. Knight, Jr., J. M. Bostick, J.R. Woodward and J. Steadman, *J. Chem. Phys.* (1983), **78**, 6415.

[12] L. B. Knight, Jr., E. Earl, A. R. Ligon and D. P. Cobranchi, *J. Chem. Phys.* (1986), **85**, 1228.

[13] A. Hasegawa, S. Kaminaka, T. Wakabayashi, M. Hayashi and M. C. R. Symons, *J. Chem. Soc., Chem. Commun.* (1983) 1199.

[14] J. R. Morton and K. F. Preston, *Mol. Phys.* (1975) **30**, 1213.

[15] D. N. Ramakrishna Rao and M. C. R. Symons, *J. Chem. Soc., Dalton Trans.* (1983), 2533.

[16] K. Toriyama, K. Nunome and M. Iwasaki, *J. Phys. Chem.* (1981), **85**, 2149.

[17] L. B. Knight, Jr., J. Steadman, D. Feller and E. R. Davidson, *J. Am. Chem. Soc.* (1984), **106**, 3700.

[18] K. Ushida, T. Shida and J. C. Walton, *J. Am. Chem.Soc.* (1986), **108**, 2805.

[19] L. B. Knight, Jr., R. L. Martin and E. R. Davidson, *J. Chem. Phys.* (1979), **71**, 3991.

[20] L. B. Knight, Jr., E. Earl, A. R. Ligon, D. P. Cobranchi, J. R. Woodward, J. M. Bostick, E. R. Davidson and D. Feller, *J. Am. Chem. Soc.* (1986), **106**, 5065.

[21] L. B. Knight, Jr., A. R. Ligon, R. W. Woodward, D. Feller and E. R. Davidson, *J. Am. Chem. Soc.* (1985), **107**, 2857.

[22] R. S. Friedman, B. J. Kelsall and L. Andrews, *J. Phys. Chem.* (1984), **88**, 1944.

[23] V. E. Bondybey, T. A. Miller and J. H. English, *J. Chem. Phys.* (1980), **72**, 2193; V. E. Bondybey, T. J. Sears, J. H. English and T. A. Miller, *J. Chem. Phys.* (1980), **73**, 2063.

[24] V. E. Bondybey, J. H. English and T. A. Miller, *Chem. Phys. Lett.* (1979), **66**, 165.

[25] L. B. Knight, Jr., J. Steadman, P. K. Miller, D. E. Bowman, E. R. Davidson and D. Feller, *J. Chem. Phys.* (1984), **80**, 4593.

[26] L. B. Knight, Jr., K. D. Johannessen, D. C. Cobranchi, E. A. Earl, D. Feller and E. R. Davidson, *J. Chem. Phys.* (1987), **87**, 885.

[27] L. B. Knight, Jr. and J. Steadman, *J.Chem. Phys.* (1982), **77**, 1750.

[28] R. S. Friedman and L. Andrews, *J. Am. Chem. Soc.* (1985), **107**, 822.

[29a] L. Andrews and B. W. Keelan, *J. Am. Chem. Soc.* (1980), **102**, 5732.

[29b]  B. J. Kelsall and L. Andrews, *J. Am. Chem. Soc.* (1983), **105,** 1413.

[30]  M. A. Baldwin, F. W. McLafferty and D. M. Jerina, *J. Am. Chem. Soc.* (1975), **97,** 6169.

[31]  J. C. Knights, J. P. M. Schmitt, J. Perrin and G. Guelachvili, *J. Chem. Phys.* (1982), **76,** 3414.

[32]  R. Robertson, D. Hils, H. Chatham and A. Gallagher, *Appl. Phys. Lett.* (1983), **43,** 544.

[33]  A. Lloret and L. Abouaf-Marguin, *Chem. Phys.* (1986), **107,** 139.

[34a]  A. Stock, A. Brandt and H. Fischer, *Chem. Ber.* (1925), **58,** 643.

[34b]  G. Urry, T. Wartik, R. E. Moore and H. I. Schlesinger, *J. Am. Chem. Soc.* (1954), **76,** 5293.

[35]  A. G. Briggs, M. S. Reason and A. G. Massey, *J. Inorg. Nucl. Chem.* (1975), **37,** 313.

# 9

# Routes to inorganic materials

## 9.1 INTRODUCTION

In recent years there has been increasing interest in the growth of thin, epitaxial layers of inorganic materials, for use in the electronics industry. These epitaxial layers may be grown from solution (liquid phase epitaxy) or, more commonly, from the gas phase by chemical vapour transport, molecular beam epitaxy or chemical vapour deposition. The last of these approaches, in which volatile organometallic or inorganic precursors are decomposed onto a substrate, has received much attention over the last 15 years.

Naturally there is speculation concerning the mechanisms by which these processes take place. Moreover, a fuller understanding of reaction mechanisms may well allow the development of more efficient processes. So there is a need to gain mechanistic information and this knowledge will come from direct observation and characterization of short-lived intermediates and from kinetic measurements. Some progress has already been made, and this chapter outlines the results of these experiments.

## 9.2 SEMICONDUCTORS

The use of semiconductors in the electronics industry dates back to the discovery of the transistor effect in germanium at the Bell Telephone Laboratories, New Jersey, in 1947. In a transistor, junctions are formed between an $n$-type semiconductor (i.e. one with an excess of electrons in the conduction band) and a $p$-type semiconductor (i.e. one with a small deficiency of electrons in the valence band). If an $n$-type sample is joined to a $p$-type sample, a $p$–$n$ junction is formed which will act as a diode for rectifying alternating current. A transistor consists of an $n$–$p$–$n$ junction, i.e. two layers of $n$-type semiconductor separated by a layer of $p$-type semiconductor. It is thus a triode, and acts as an amplifier.

The group IV elements germanium and silicon are examples of semiconductors, and both have been used widely in the electronics industry. They may be converted to *n*-type semiconductors by doping with a group V element such as arsenic, or into *p*-type semiconductors by doping with a group III element such as boron. One approach to the formation of thin layers of silicon is to decompose silanes, either thermally or by the action of a microwave discharge. Important intermediates in these reactions are the divalent silylene species, and Chapter 5 describes some of the methods used for the production and characterization of these intermediates.

Other materials which act as semiconductors are the binary compounds of group III and group V or group II and group VI elements. Thin layers of these materials may be grown by metal organic chemical vapour deposition (MOCVD), where mixtures of organometallic and inorganic precursors are thermally decomposed. Table 9.1 lists some of the compounds which have been grown in this way.

**Table 9.1** — MOCVD of some inorganic compounds

| Compound | Reactants | Growth temperature/°C |
|---|---|---|
| GaAs | $GaMe_3 + AsH_3$ | 600–750 |
| GaP | $GaMe_3 + PH_3$ | 700–800 |
| GaSb | $GaMe_3 + SbMe_3$ | 500–550 |
| AlAs | $AlMe_3 + AsH_3$ | 700 |
| AlN | $AlMe_3 + NH_3$ | 1250 |
| GaN | $GaMe_3 + NH_3$ | 925–975 |
| GaN | $GaEt_3 + NH_3$ | 800 |
| InP | $InEt_3 + AsH_3$ | 650–700 |
| InSb | $InEt_3 + SbEt_3$ | 460–475 |
| ZnS | $ZnEt_2 + H_2S$ | 750 |
| ZnSe | $ZnEt_2 + H_2Se$ | 725–750 |
| ZnTe | $ZnEt_2 + TeMe_2$ | 500 |
| CdS | $CdMe_2 + H_2S$ | 475 |
| CdSe | $CdMe_2 + H_2Se$ | 500 |
| CdTe | $CdMe_2 + TeEt_2$ | 400 |

The first inorganic material to be produced by MOCVD was GaAs, which was made by decomposing a mixture of trimethyl gallium and arsine in a stream of hydrogen carrier gas on a substrate heated to 600–800°C [1]. Soon, however, the technique was extended to a wide range of semiconductor compounds, largely as a result of the pioneering work of Manasevit. There are two questions concerning the mechanisms of these reactions which are of particular importance. First, are adducts formed between the two precursor molecules in the gas phase prior to decomposition? Second, what is the role played by free radicals in the process?

In answer to the first question, it is clear that adducts are formed during the growth of III–V semiconductors such as GaAs and InP. Indeed some of these adducts, e.g. $(CH_3)_3In.P(C_2H_5)_3$, may be presynthesized and transported, and hence used directly in the MOCVD process [2]. However, adduct formation may also pose problems. For example, the unstable adduct $(CH_3)_3In.PH_3$ decomposes, on heating, to yield an involatile polymer rather than the desired InP [3]:

$$In(CH_3)_3 + PH_3 \longrightarrow (CH_3)_3In.PH_3 \longrightarrow \{In(CH_3)PH\}_n + 2nCH_4$$

Adducts formed between group II and group VI organometallic precursors appear to be unstable, and their possible rôle in II–VI semiconductor formation is still uncertain. It has been claimed that the adduct $(CH_3)_2Cd.Te(C_2H_5)_2$ is formed during the thermal growth of CdTe from mixtures of $(CH_3)_2Cd$ and $(C_2H_5)_2Te$ [4]. However, further experiments are required to characterize this species unequivocally.

There has been a recent development of the MOCVD process in which lower temperatures are employed, and photolysis is used to assist the deposition process. This approach, which is often known as photo-epitaxy, has been used successfully to grow, using an apparatus similar to that shown in Fig. 9.1, epitaxial layers of HgTe from mixtures of $(C_2H_5)_2Te$ and Hg vapour in $H_2$ or He carrier gas. A substrate temperature of only 200°C was required for epitaxial growth — a reduction of 200°C from that required for the conventional thermal deposition of this material.

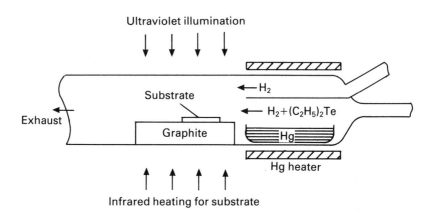

Fig. 9.1 — Schematic view of a reactor for growing thin layers of HgTe by chemical vapour deposition. Reproduced with permission from Irvine and Mullin, *Chemtronics*, **3**, 54 © 1987 Butterworths.

It is probable that the photolytic method, like the thermal route, proceeds via a free radical mechanism. To some extent, information about free radical intermediates can be obtained by analysing the stable gas phase reaction products, one method of analysis being mass spectrometry. Thus it has been shown that pyrolysis or photolysis of $(C_2H_5)_2Te$ in an atmosphere of He or $H_2$ yields mixtures of ethane, ethene and butane [5]. It is proposed [5] that ethene may be formed from an ethyl radical:

$$C_2H_5^{\bullet} \longrightarrow C_2H_4 + H^{\bullet}$$

while a possible route to ethane involves reaction of a hydrogen atom with the relatively stable radical $C_2H_5Te^{\bullet}$, formed on decomposition of $(C_2H_5)_2Te$:

$$C_2H_5Te_2^{\bullet +C}H_5^{\bullet}$$

$$C_2H_5Te^{\bullet} + H^{\bullet} \longrightarrow C_2H_6 + Te$$

It is most likely that butane is formed by a recombination reaction such as

$$C_2H_{5+}^{\bullet}C_2H_5^{\bullet} \longrightarrow C_4H_{10}$$

or

$$C_2H_5^{\bullet} + C_2H_5Te^{\bullet} \longrightarrow C_4H_{10} + Te$$

As yet most of these mechanisms are conjectural. Indeed there is no direct evidence for the intermediacy of the radical $C_2H_5Te^{\bullet}$, although it is known that some relatively long-lived intermediate is involved. However, gas phase pyrolysis of $(CH_3)_2Hg$ yields $CH_3Hg^{\bullet}$, and this species has been trapped in an inert gas matrix and characterized by infrared spectroscopy [6]. Similar experiments could well lead to the direct observation $C_2H_5Te^{\bullet}$.

## 9.3   METAL NITRIDES AND OXIDES

Metal nitrides and oxides are inorganic materials which find use in the electronics industry as wide-band gap semiconductors, or as insulating layers. There is much interest in methods by which thin layers of these materials may be grown from volatile precursors.

The wide band-gap semiconductor gallium nitride has been produced by thermal decomposition, at 1000°C, of mixtures of $(CH_3)_3Ga$ and $NH_3$ on a sapphire substrate [7].

$$(CH_3)_3Ga + NH_3 \xrightarrow{1000°C} GaN + 3CH_4$$

It is known that the adduct $(CH_3)_3Ga.NH_3$ may be produced by reaction of $(CH_3)_3Ga$ and $NH_3$ at room temperature. It therefore seems likely that similar adducts of the type $(CH_3)_nGa.NH_3$ are formed during the deposition of GaN, where $n$ decreases as the temperature is increased. Boron nitride may be grown, similarly, from the adducts $(C_2H_5)_3N.BH_3$ or $(C_2H_5)_2HN.BH_3$ at temperatures in the range 450–800°C [8]. Analysis of the BN product from this reaction shows that the B:N ratio is not 1:1. In fact there is a large excess of boron (as high as 10:1 in some samples) which suggests that decomposition of the adduct, and loss of N-containing species occurs during the reaction. Infrared spectroscopic analysis of the BN product also shows the presence of B–H and N–H moieties, especially in samples grown at lower temperatures [8]. This observation, too, has mechanistic implications, suggesting that decomposition of the adduct is only complete at high temperatures.

An alternative route to metal nitrides — but one which leads to the possible incorporation of halide impurity in the nitride sample — is to pass a mixture of a volatile metal halide and nitrogen through a microwave discharge. In this way, for example, titanium nitride may be formed from $TiCl_4$ and it is interesting to note that the TiN monomer has been trapped and characterized in inert gas matrices, following production in this way [9].

Metal oxides have been generated by thermal decomposition of volatile metal complexes with oxygen-containing ligands. Suitable precursors include metal alkoxide and acetylacetonate complexes [10], and a specific example is provided by the chromium actylacetonate complex $Cr(acac)_3$ (**1**). Presumably such complexes decompose via loss of organic radicals, but as yet there is little concrete evidence

(**1**)

regarding the mechanisms of these reactions. An alternative route to metal oxides is via the photochemical reaction of metal carbonyl molecules with oxygen. These reactions have been the subjects of some quite detailed matrix-isolation investigations, and have already been described in section 3.9.


## 9.4  METAL SILICIDES

Metal silicides, which, depending on their composition, may be electrical conductors or semiconductors, form an essential part of silicon-based semiconductor technology. One of their main uses is to provide a barrier, between a silicon substrate and an attached metal conductor, which inhibits diffusion of silicon or metal atoms between the layers. One route to metal silicides is the plasma reaction — for example, in a microwave discharge cavity — of a volatile metal halide with a silane. Tungsten silicide has been generated, in this way, from tungsten hexafluoride:

$$WF_6 + SiH_4 \xrightarrow[\text{plasma}]{570\,K} WSi_2 + \text{other products}$$

However, halide contamination of the final product remains a limitation of this process, so a search has been made for suitable organometallic precursors.

Some work has been carried out on volatile organometallic compounds which contain a metal silicon bond. To a large extent, this work has concentrated on the silyl metal carbonyl derivatives, or general formula $R_3SiM(CO)_n$, where M is a transition metal and R is either hydrogen or an alkyl group. Such precursors may be decomposed onto a variety of substrates — including Si, $SiO_2$, GaAs and metals — at temperatures in the range 670–770 K, yielding metal silicides, as shown in Table 9.2 [11].

**Table 9.2** — Precursors and products in metal silicide chemical vapour deposition

| Precursor | Product |
|---|---|
| $H_3SiCo(CO)_4$ | CoSi |
| $H_3SiMn(CO)_5$ | $Mn_5Si_3 + MnSi_x$ $(x \sim 1.25)$ |
| $H_3SiRe(CO)_5$ | $Re_5Si_3 + \ldots$ |
| $(H_3Si)_3Fe(CO)_4$ | $\beta$-$FeSi_2$ |

Mass spectrometric studies on these reactions have suggested that various decomposition pathways might exist. One is a single stripping reaction in which $H_2$ and CO are removed:

$$H_3SiCo(CO)_4 \longrightarrow H_mSiCo(CO)_n + H_2 + CO$$

However, another route involves oxygen atom migration to silicon, a process which has been observed in the controlled pyrolysis of $Me_3SiCo(CO)_4$.

$$Me_3SiCo(CO)_4 \xrightarrow[50\,h]{380\,K} Me_3SiOCCo_3(CO)_9 + (Me_3SiOC)_4\,Co(CO)_4 + \ldots$$

Thus silicon oxide and metal carbide phases may be introduced into the final silicide product, although this problem may be avoided by careful control of the thermolysis conditions during the CVD reaction. It is possible that a photochemical reaction would lead to a more controlled decomposition. More work is required here to elucidate the reaction mechanisms, since this knowledge would help to find the optimum conditions for pure metal silicide deposition.

## 9.5  LASER-WRITING EXPERIMENTS

Photochemical laser writing is a variation on the photo-epitaxy technique, where an ultraviolet laser is used to initiate deposition of a metal by photo-disscociation of an organometallic precursor close to, or adsorbed onto, a substrate surface. Typical examples include depostion of Al from $(CH_3)_3Al$, Zn from $(CH_3)_2Zn$ and W from $W(CO)_6$ [12]. By using a highly focused laser beam it is possible to deposit a thin line of metal whose width is less than 1 $\mu$m. This approach has important applications in the deposition of conducting links in integrated circuit patterns. The mechanisms of such processes are, however, complicated. In the first place, it is often difficult to distinguish between photochemical and thermal decomposition when lasers are used on precursors adsorbed onto substrates. Secondly, kinetic studies of the thermal and photochemical decomposition of metal alkyls such as $(CH_3)_2Zn$, $(CH_3)_2Cd$ and

$(CH_3)_3Al$, which were carried out in the 1950s and 1960s, revealed that the gas phase processes are often complicated by surface reactions. Surface effects are of particular importance when a thin layer of metal is being grown on a substrate. An individual metal alkyl may pose its own mechanistic problems. For example $(CH_3)_3Al$ is likely to be partially dimerized under the experimental conditions commonly used for laser writing of Al. Thus $(CH_3)_6Al_2$, rather than $(CH_3)_3Al$, may well be the more important photochemical precursor in these experiments.

A process which is closely related to that discussed above is the use of a low-power ultraviolet laser to initiate localized photochemical etching of a semiconductor surface [13]. In these experiments, the laser light is used to photodissociate an alkyl halide molecule, such as $CH_3Br$, close to the semiconductor surface. The halogen atoms so produced are believed to be adsorbed onto the surface, which is eroded by subsequent vaporization of the volatile halide products:

$$CH_3Br_{(g)} \xrightarrow{h\nu} CH_{3(g)} + Br_{(g)}$$

$$Br_{(g)} + GaAs_{(s)} \longrightarrow GaAs\ldots Br_n \text{ (ads.)}$$

$$GaAs\ldots Br_n \text{ (ads.)} \longrightarrow GaBr_{x(g)} + AsBr_{y(g)}$$

Although there is much concrete knowledge concerning the gas phase photolysis of metal carbonyls (see section 3.14) and of metal alkyls [14], most of the mechanisms proposed for these laser-initiated reactions are uncertain. As with other examples discussed in this chapter, a fuller understanding of reaction mechanisms would be important in the development of these processes.

## 9.6   POLYTHIAZYL

This chapter ends with a discussion of routes by which polymeric sulphur nitride, $(SN)_x$ — a material often referred to as 'polythiazyl' — may be prepared. Since 1973, when Labes et al. suggested that polythiazyl is metallic [15], there has been intense interest in this inorganic material. Subsequent studies [16] showed that it remains metallic down to very low temperatures, becoming superconducting near 0.3 K. Furthermore, oriented epitaxial films of $(SN)_x$, grown on polymer substrates, show a high optical anisotropy throughout the near-infrared and low-frequency portion of the visible spectrum, suggesting possible applications in optical devices.

The usual routes to 'polythiazyl' employ the cage complex $S_4N_4$ (**2**) as the starting material. As long ago as 1910, Burt [17] demonstrated that when $S_4N_4$ vapour was passed over heated silver or quartz wool in a vacuum system, an involatile deposit was produced which appears blue when viewed by transmitted light, but by reflected light has a bronze colour with a metallic lustre. On the basis of elemental analysis he stated that this deposit was a polymeric species of formula $(SN)_x$ — almost certainly polythiazyl — and suggested that it was formed from $S_4N_4$ via a more volatile, unstable sulphur nitride. Subsequent experiments have been performed in which the gaseous intermediates, produced by passage of $S_4N_4$ vapour over heated silver wool, are trapped in a low-temperature matrix and are studied by infrared and Raman spectroscopy [18]. These experiments show that the principal intermediate is the cyclic molecule $S_2N_2$ (**3**). Fig. 9.2 shows the Raman spectrum of matrix-isolated $S_2N_2$

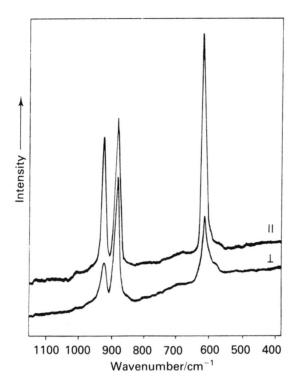

Fig. 9.2 — Raman spectra of $S_2N_2$ in an argon matrix recorded with the polarizer in parallel ($\parallel$) and perpendicular ($\perp$) positions. Reproduced with permission from Downs and Hawkins, *Adv. Infrared Raman Spectrosc.*, **10**, 1 © 1983 John Wiley.

produced in this way; the bands arising from vibrations of $a_1$ symmetry can clearly be assigned from the polarization measurements. Thus it is clear that $S_2N_2$ is an important intermediate in the formation of 'polythiazyl', and these recent matrix-isolation experiments nicely confirm and explain the early findings of Burt.

                    (2)                                              (3)

While it is known that crystals of $S_2N_2$ will slowly polymerize at room tempera-ture to give 'polythiazyl' [19], the mechanism of this process remains unclear. It is possible that the reaction is initiated by some other species — possibly the thiazyl monomer NS•. Recently it has been shown that, alongside $S_2N_2$, red and brown volatile materials are produced when $S_4N_4$ vapour is cracked over a silver or a silver selenide catalyst. On the basis of elemental analysis, ESR and mass spectrometric measurements, it has been claimed that this red material consists almost entirely of NS• [20]. However, matrix isolation of the vapour species produced by the passage of $S_4N_4$ vapour over heated silver selenide gives unequivocal infrared evidence not of NS•, but of thionylimide, HNSO, in addition to $S_2N_2$ [21]. Fig. 9.3 illustrates part of the IR spectrum of matrix-isolated HNSO, generated from $S_4N_4$ partially enriched in

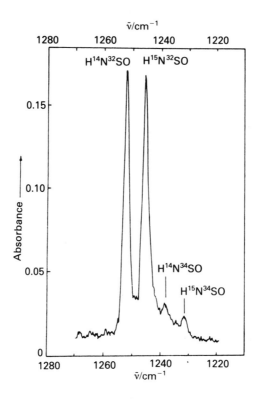

Fig. 9.3 — Infrared spectrum of HNSO in an argon matrix produced by passing $S_4N_4$ vapour 50% enriched in $^{15}N$ over an $Ag_2Se$ catalyst at 160°C. Reproduced with permission from Almond *et al.*, *Polyhedron*, **7**, 629 © 1988 Pergamon.

$^{15}N$. The observed isotope splitting patterns leave no doubt about the assignment of this feature to $v_2$ of the HNSO molecule. It is possible, therefore, that HNSO is an active intermediate in some 'polythiazyl' preparations, and this finding may account for the observation of hydride impurity (typically 1–4%) in some samples of the polymer. Of particular significance, in this respect, is the observation that the hydride impurity may be increased to about 8% if water vapour is present during the polymerization process.

## 9.7 SUMMARY

The aim of this chapter has been to provide a short account of areas where a knowledge of short-lived intermediates has helped to understand some preparative routes to inorganic materials. This is an important area of chemistry and one which crosses the traditional boundaries between chemistry and other scientific disciplines such as physics and materials science. The development and exploitation of new inorganic materials will require considerable work from scientists whose interests span many different areas. But the mechanistic chemist will have a distinct role to play. It is only by understanding the mechanisms of the preparative reactions that more efficient processes may be developed. Furthermore, this knowledge will almost certainly help in the extension of these preparative routes to a wide range of new materials.

## REFERENCES

[1] H. M. Manasevit, *Appl. Phys. Lett.* (1968), **12**, 156.
[2] H. Renz, J. Weidlein, K. W. Benz and M. Pilkuhn, *Electron. Lett.* (1980), **16**, 228; R. H. Moss and J. S. Evans, *J. Crystal Growth* (1981), **55**, 129.
[3] R. Didchenko, J. E. Alix and R. H. Toeniskoetter, *J. Inorg. Nucl. Chem.* (1960), **14**, 35.
[4] J. B. Mullin, S. J. C. Irvine and D. J. Ashen, *J. Crystal Growth* (1981), **55**, 92.
[5] S. J. C. Irvine and J. B. Mullin, *J. Crystal Growth* (1986), **79**, 371.
[6] A. Snelson, *J. Phys. Chem.* (1970), **74**, 537.
[7] M. Hashimoto, H. Amano, N. Sawaki and I. Akasaki, *J. Crystal Growth* (1984), **68**, 163.
[8] D. M. Schleich, W. Y. F. Lai and A. Lam, in *Transformation of Organometallics into Common and Exotic Materials*: *Design and Activation*, p. 178, R. M. Laine (ed.), Martinus Nijhoff: Dordrecht (1988).
[9] F. W. Froben and F. Rogge, *Chem. Phys. Lett.* (1981), **78**, 264.
[10] D. C. Bradley, *New Scientist* (1988), 118—1608, 38.
[11] B. J. Aylett in *Transformation of Organometallics into Common and Exotic Materials*: *Design and Activation*, p. 165, R. M. Laine (ed.), Martinus Nijhoff: Dordrecht (1988).
[12] D. J. Ehrlich, R. M. Osgood, Jr. and T. F. Deutsch, *J. Vac. Sci. Technol.* (1982), **21**, 23.
[13] D. J. Ehrlich, R. M. Osgood, Jr. and T. F. Deutsch, *Appl. Phys. Lett.* (1980), **36**, 698.
[14] C. F. Yu, F. Youngs, K. Tsukiyama, R. Bersohn and J. Preses, *J. Chem. Phys.* (1986), **85**, 1382.
[15] V. V. Walatka, Jr., M. M. Labes and J. H. Perlstein, *Phys. Rev. Lett.* (1973), **31**, 1139.
[16] R. L. Greene, G. B. Street and L. J. Suter, *Phys. Rev. Lett.* (1975), **34**, 577; A. A. Bright, M. J. Cohen, A. F. Garito, A. J. Heeger, C. M. Mikulski and A. G. MacDiarmid, *Appl. Phys. Lett.* (1975), **26**, 612.
[17] F. P. Burt, *J. Chem. Soc.* (1910), **97**, 1171.
[18] A. J. Downs and M. Hawkins, *Adv. Infrared Raman Spectrosc.* (1983), **10**, 1; R. Evans, D. Phil. thesis, University of Oxford, 1980.

[19] M. Goehring and D. Voigt, *Z. Anorg. Allg. Chem.* (1956), **285**, 181.
[20] P. Love, G. Myer, H. I. Kao, M. M. Labes, W. R. Junken and C. Elbaum, *Ann. New York Acad. Sci.* (1978), **313**, 745.
[21] M. J. Almond, A. J. Downs and T. L. Jeffery, *Polyhedron* (1988), **7**, 629.

# 10

# Atmospheric and interstellar chemistry

## 10.1 INTRODUCTION

Some of the most interesting experiments performed on short-lived molecules are those which have helped to understand the chemical processes taking place in the earth's atmosphere and in interstellar space. In the first place, spectroscopic characterization of several short-lived molecules in the laboratory has allowed recognition of the same molecules in outer space. For example, the discovery of the HNC molecule as an interstellar component was only possible because of the availability of spectroscopic data for HNC from the matrix-isolation experiments of Milligan and Jacox [1a]. Following the observation of HNC in interstellar space, the assignment was confirmed by the laboratory measurement of its microwave spectrum [1b]. Matrix isolation has proved to be a successful means of characterizing other short-lived astronomical molecules. Species characterized in this way include radicals such as the ethynyl radical ($\cdot\,C_2H$) found in the Orion nebula, 'high-temperature' diatomic molecules such as TiO and FeH, and larger organic molecules such as the substituted polyacetylene $HC{\equiv}CC{\equiv}CC{\equiv}CC{\equiv}N$.

Laboratory modelling of chemical reactions has helped to interpret the processes taking place in the earth's atmosphere. Thus matrix-isolation experiments have been used to investigate the formation of the chlorine nitrate molecule ($ClONO_2$) from the radicals ClO and $NO_2$, a process which is important in the destruction of stratospheric ozone by chlorofluorocarbons ('freons'). The scope of matrix isolation has been further increased by its use as a technique for trace gas analysis on atmospheric samples.

## 10.2 CHLORINE NITRATE, AND ATMOSPHERIC OZONE DEPLETION

A topical issue of great environmental concern is that of the depletion of ozone in the stratosphere. Two radicals which are believed to be active in this depletion are $NO_2$ and ClO, the latter being derived, at least partially, from chlorofluorocarbons ('freons') released into the atmosphere. Thus the combination reaction

$$ClO + NO_2 \longrightarrow ClONO_2$$

is of importance because it reduces the concentrations of these radicals. Chlorine nitrate, $ClONO_2$, is photolysed only slowly by ultraviolet light to regenerate ClO and $NO_2$, so it acts as a temporary reservoir. However, the rate of decay of ClO in the gas phase, as measured in the laboratory, appears to show some anomalies which can be interpreted by assuming that a second, more photolabile isomer of $ClONO_2$ — possibly ClOONO — is produced. If this premise is correct, the efficiency of the combination reaction in acting as a reservoir for ClO and $NO_2$ would be reduced, and the calculated depletion of stratospheric ozone by 'freon' release increased. Thus a detailed investigation of the combination reaction is of some importance.

Burrows, Griffith, Schuster and others at the Max Planck Institute in Mainz injected gaseous ClO and $NO_2$ into a stirred-flow photolytic reactor [2] and they have analysed the outcome of the action in one of two ways: first, by measuring the ultraviolet spectrum of the gaseous sample, and second by freezing the gaseous reaction products with a large excess of nitrogen carrier gas, and recording the infrared spectrum of the condensate.

The latter approach is particularly rewarding. The infrared spectrum can be used to assay quantitatively the major reaction products, and to search for any minor products. It is also possible to carry out matrix isolation of both unphotolysed and photolysed gaseous mixtures. Thus the effect of photolysis may be explored.

In this way it has been shown that, at a pressure of 3000 $Nm^{-2}$ and at temperatures in the range 253–298 K, $ClONO_2$ is the sole detectable product of the combination reaction of ClO and $NO_2$, being formed in a yield of 93±2%. The results seem to put an end to speculation about the formation of chlorine nitrate isomers. Moreover they make out a strong case for using matrix isolation on other systems to supplement the methods more traditionally associated with studies of the kinetics and mechanisms of gas phase reactions.

## 10.3  ATMOSPHERIC TRACE GAS ANALYSIS

Matrix isolation can be used alongside a wide variety of other methods for atmospheric trace gas analysis. This approach, which has been developed by Griffith and Schuster, involves three stages: first, cryogenic sampling of the air; second, the growth of a carbon dioxide matrix from the sample; finally infrared spectroscopic analysis of the matrix sample. Several advantageous features are associated with this technique. In particular it has a wide applicability, and the use of infrared spectroscopy confers good selectivity so that many species can be sampled and measured simultaneously. Furthermore a high sensitivity (about 1–10 parts in $10^{12}$), precision and accuracy are obtained.

The sampling procedure is as follows. An oil-free membrane pump is used to draw the air to be analysed through a liquid-nitrogen-cooled glass sampler fitted with teflon stopcocks at a rate of about 10 l $min^{-1}$. Thus all components which are condensable at 77 K are retained; $N_2$, $O_2$, Ar, $H_2$, $CH_4$ and CO escape collection, while $O_3$ and NO are probably not trapped quantitatively since both have significant vapour pressures at 77 K. The major collected component from the lower atmosphere is water, whereas from the upper atmosphere it is carbon dioxide. Following collection in the field, the sample is stored under liquid nitrogen, and returned to the laboratory. The presence of water is a major problem because its own infrared spectrum is intense and broad and it tends to have a broadening effect on the absorptions of other molecules, thus decreasing the selectivity of the technique. To remove water vapour, the sample is warmed to 220–230 K so as to allow carbon dioxide and trace gases to sublime, while the water is retained. The gaseous sample is then frozen rapidly — typically in 1 min via a stainless-steel valve — onto a gold mirror surface coating a copper cold finger of a liquid-nitrogen-cooled cryostat. The matrix so produced is typically ca. 4 mm in diameter and ca. 300 $\mu$m thick at its centre, and is composed principally of carbon dioxide. In a typical experiment the next most abundant constituents are $N_2O$ and residual water.

This technique has been used to investigate, for example, air samples collected at Schauinsland in the Black Forest in Germany [3]. By using calibration experiments involving gaseous mixtures of known composition which were prepared in the laboratory, it proved possible to analyse these samples quantitatively for $N_2O$, $CFCl_3$, $CF_2Cl_2$, OCS, $CS_2$ and PAN (peroxyacetylnitrate). Other features — some identified, others unidentified — were seen. Fig. 10.1 shows the region 1600–1850 $cm^{-1}$ for two of the samples in which various contaminants have been identified

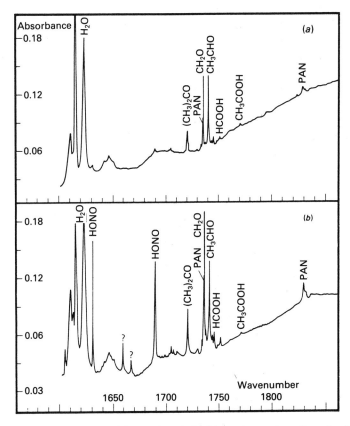

Fig. 10.1 — Infrared spectra of $CO_2$ matrices derived from air samples collected at Schauinsland in the Black Forest, Germany on 15 February 1984: (a) at 8.00 a.m.; (b) at 5.00 p.m. Reproduced with permission from Griffith and Schuster, *J. Atmospheric Chem.*, **5**, 59 ©1987 D. Reidel.

qualitatively. The presence of various organic carbonyl compounds — e.g. $H_2CO$, HCOOH, $CH_3CHO$, $CH_3COOH$ and $(CH_3)_2CO$ — can clearly be seen alongside oxynitrogen compounds, e.g. HONO. The spectra show how much dirtier the air has become during the course of a day. In particular there is a dramatic increase in the concentration of HONO between 8.00 a.m. and 5.00 p.m., presumably as a result of car exhaust fumes.

The main drawback of this technique is the problem of sampling. The three-stage procedure — collecting, drying, freezing — makes the method slow, and further work is required to ensure that the chemical composition of the frozen samples accurately reflects the composition of the air from which they were originally collected.

## 10.4 INTERSTELLAR GRAINS
Within the interstellar medium are found small interstellar grains. These are believed to comprise silicate particles with diameters in the order of 0.05 $\mu$m, which are the nuclei for frozen mantles of condensable molecules typically about 0.05 $\mu$m thick. Such grains are normally at temperatures near 10 K. Fig. 10.2 depicts three types of grain which vary according to the stage of their life cycle. Stage (a) represents a diffuse cloud grain which is coated with highly processed, refractory organic material. Stage (b) represents a grain which has grown in a molecular cloud where the intensity of ultraviolet radiation is quite high. Under these conditions the refractory organic mantle is coated with an outer layer of so-called 'dirty ice', which contains several molecules derived from the molecular cloud such as $H_2O$, $CH_4$ and $NH_3$. Prolonged exposure to ultraviolet and vacuum ultraviolet radiation causes this outer mantle to resemble a photolysed matrix, and it is expected to be host to a variety of photoproducts, for example the radicals HCO•, •$NH_2$ and HOCO•. Stage (c) represents a grain which has grown in a dense molecular cloud close to a

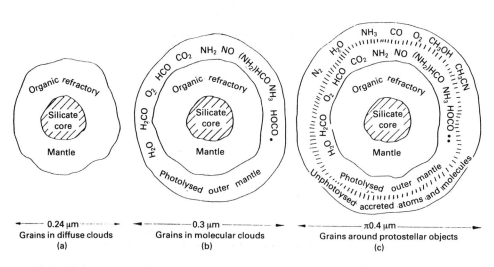

Fig. 10.2 — Schematic view of different types of grain expected in different regions of the interstellar medium. Reproduced with permission from Greenberg *et al.*, *J. Phys. Chem.*, **87**, 4243 ©1983 American Chemical Society.

protostellar object. Here photolysis has essentially stopped and the dirty ice layer is capped by an outermost mantle of 'ice' containing molecules such as $N_2$, $H_2O$, $NH_3$ and CO, which have not been processed by irradiation.

These various types of grain are sufficiently different from one another that they may be distinguished by characteristic features in their infrared spectra. The problem facing the astronomical spectroscopist is to interpret these spectra, and from them to gain some knowledge of the species present within the grain mantles.

One approach which has been developed at the University of Leiden in Holland and at the NASA Research Center at Ames in California, by Allamandola, Baas, Greenberg, and their co-workers, is to simulate, in the laboratory, the essential conditions in interstellar space as they affect grain formation [4,5]. A schematic representation of such an experiment is shown in Fig. 10.3. Here, interstellar grain formation is modelled by depositing a gaseous mixture onto a cold finger (an aluminium block for reflectance or a caesium iodide window for transmission spectroscopy), which fulfils the role of silicate core of a real interstellar grain. The gaseous mixture consists of a mixture of simple molecules such as $CH_4$, CO, $CO_2$, $H_2O$, $NH_3$ and $CH_3OH$, in which the elemental ratio O:C:N approximates to their

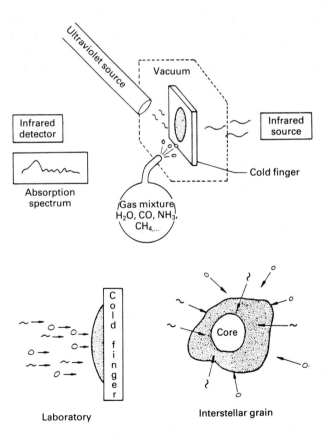

Fig. 10.3 — The laboratory analogue method for interstellar grain evolution. Reproduced with permission from Greenberg *et al.*, *J. Phys. Chem.*, **87**, 4243 ©1983 American Chemical Society.

cosmic abundance of 6:3:1. The rate of deposition of the mixture is adjusted to give a growth rate of the thickness of the condensate of ca. 6 $\mu$m h$^{-1}$, and in a typical experiment the final 'matrix' ia about 1.0 $\mu$m thick.

The gaseous mixture is subjected to vacuum ultraviolet irradiation either during or following deposition. The usual vacuum ultraviolet source is a microwave-excited hydrogen lamp which has a sharp emission peak at 121.6 nm and a second broad emission spanning about 50 nm and centred near 160 nm. This spectrum represents quite well the photolysis conditions prevalent in a diffuse interstellar cloud. Infrared spectroscopy is used to monitor the response of the sample at different stages of the experiment. One major difference between the natural process occurring in inter-stellar space and the laboratory experiment is the timescale; an experiment of a few hours' duration is being used to model a process in space which takes place over a period of several thousand years!

An example of laboratory simulation is provided by an experiment in which a mixture of CO and $NH_3$ in the proportion 3:1 has been frozen at 10 K and irradiated with vacuum ultraviolet light [6]. The photolysed ice has been warmed to 150 K so as to allow the more volatile components to evaporate, and the residue then recooled to 10 K. Prior to photolysis the infrared absorption spectrum of the ice shows only a single band in the region 2100–2200 cm$^{-1}$, and this can be assigned to CO. As shown in Fig. 10.4 there is a conspicuous change following photolysis, and partial evapo-ration of the ice. A weak absorption at 2140 cm$^{-1}$, presumably due to CO trapped in the solid residue, is now accompanied by a moderately strong absorption centred at

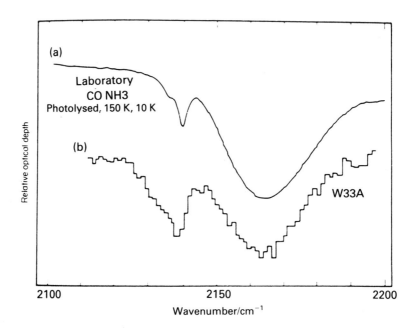

Fig. 10.4 — The laboratory spectrum of photolysed CO/NH$_3$ ice which has been warmed to 150 K then recooled to 10 K, compared with the spectrum of interstellar source W33A. Reproduced with permission from Lacy *et al.*, *Astrophys. J.*, **276**, 533 ©1984 American Astronomical Society.

2165 cm$^{-1}$. Fig. 10.4 also depicts the astronomical spectrum derived from the interstellar object W33A, one of several compact infrared sources and possibly a very young star embedded in a molecular cloud. The infrared spectra derived from W33A, and from the photolysed CO:NH$_3$ 'ice' show a striking similarity — in particular the absorption at 2165 cm$^{-1}$ is present in both. In laboratory experiments it is found that this band, at 2165 cm$^{-1}$, is generated only in ices containing carbon and nitrogen precursors. These circumstances, and the position of the new band, suggest strongly that the absorber contains a cyano group $-C\equiv N$. The obvious inference is that molecules containing similar cyano groups are also present in the ice mantles around the interstellar grains responsible for the spectrum derived from W33A.

## 10.5  DISULPHUR AND THE FORMATION OF COMETS

Recently the ultraviolet-visible emission spectrum of the comet IRAS-Araki-Alcock 1983 VII has been recorded [7]. Included in this spectrum is a strong, structured emission which is undoubtedly due to the $B^3\Sigma_u^- \to X^3\Sigma_g^-$ transition of the disulphur molecule, S$_2$. However, it is known that S$_2$, unlike some other sulphur-containing molecules such as H$_2$S, has an exceedingly low concentration in interstellar space. Thus it is unlikely that S$_2$ owes its presence in the comet simply to condensation from the interstellar medium.

A more plausible explanation is that S$_2$ is formed by ultraviolet photolysis of more abundant sulphur molecules, e.g. H$_2$S, in the 'dirty ice' mantle which condenses around interstellar grains. To investigate this hypothesis a series of laboratory experiments have been carried out, in which dirty ices including H$_2$S (e.g. H$_2$O/CO/CH$_4$/H$_2$S) have been condensed and illuminated with ultraviolet light at 12 K. Laser-excitation of the photolysed condensate at $\lambda = 308$ nm is then found to yield the emission bands characteristic of the $B^3\Sigma_u^- \to X^3\Sigma_g^-$ transition of S$_2$. In these experiments, when the sulphur-to-oxygen ratio in the starting materials is adjusted to the cosmic abundances of the two elements, the ratio [S$_2$]:[H$_2$] in the photolysed condensate is found to vary between $2\times10^{-4}$:1 and $1.4\times10^{-3}$:1. These values are consistent with the observed ratio of [S$_2$]:[H$_2$O]$=5\times10^{-4}$:1 in the comet IRAS-Araki-Alcock 1983 VII.

Thus the result of the laboratory experiments suggests that cometary S$_2$ molecules are formed photolytically in the mantles of interstellar grains. Furthermore they have implications concerning the actual formation of comets, providing support ₊or the hypothesis that comets are formed by the coagulation of small dust particles.

## 10.6  UNIDENTIFIED INFRARED EMISSION

Associated with many celestial objects (including nebulae) are unidentified infrared (UIR) emission bands and diffuse interstellar absorption bands (DIB). UIR consists typically of strong bands at ca. 3000, 2940, 1600, 1300, 1160 and 880 cm$^{-1}$ with weaker features at ca. 1800 and 1440 cm$^{-1}$ [8]. The relative intensities of these bands vary somewhat, although the bandwidths appear to be more-or-less independent of the source.

In 1981, Duley and Williams made the observation that some of the UIR band frequencies are characteristic of large polycyclic aromatic hydrocarbons (PAHs) such as coronene [9]. Allamandola and his co-workers have compared the UIR emission spectrum from the Orion bar nebula with the Raman spectrum obtained from car exhaust material. These two seemingly unrelated sources show some striking similarities [10], as illustrated by the region 1000–1900 cm$^{-1}$ of both spectra, shown in Fig. 10.5. The car exhaust material is composed of a mixture of PAHs and graphite. Thus the similarity to the interstellar spectrum lends support to the hypothesis of Duley and Williams that UIR results from interstellar PAH molecules. Allamandola has come up with the novel proposition of 'auto exhaust along the Milky Way'!

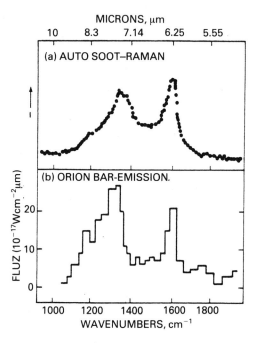

Fig. 10.5 — (a) Raman spectrum of car exhaust material; (b) unidentified infrared (UIR) emission from the Orion bar nebula. Reproduced with permission from Allamandola *et al.*, *Astrophys, J.*, **290**, L25 ©1985 American Astronomical Society.

Since PAH molecules have low ionization energies, it is likely that, at least to some extent, they exist in their cationic forms in the interstellar medium. It has been suggested that PAH cations are the carriers of the diffuse interstellar absorption bands (DIB) which occur in the visible part of the spectrum, and whose origin has remained a mystery since their discovery more than 50 years ago [11]. At present there is little spectroscopic data on gas phase PAH cations. However, visible absorption spectra have been measured for a number of PAH cations trapped in low-temperature 'freon' matrices [12]. Such highly polarizable matrices will tend to induce a significant red shift (i.e. a shift to lower frequency) of the absorption bands, with respect to the gas phase. If an estimate (of ca. 1550 cm$^{-1}$) is made for this red shift, then the laboratory spectra for PAH cations show some similarity to the diffuse interstellar bands. Some such spectra are illustrated in Fig. 10.6 [13].

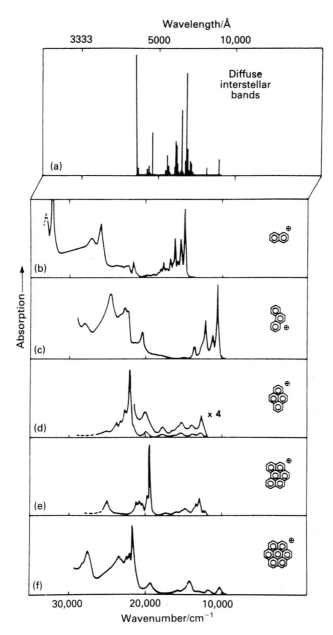

Fig. 10.6 — (a) Schematic representation of the diffuse interstellar band (DIB) spectrum; (b) to (f) visible absorption spectra of five polyaromatic cations isolated in freon matrices. These spectra have been blue-shifted by 1500 cm$^{-1}$ to compensate for the inherent matrix red shifts. Reproduced with permission from Crawford *et al.*, *Astrophys. J.*, **293**, L45 ©1985 American Astronomical Society.

The circumstances allow no more than tentative conclusions, but the comparison is promising and a plausible case can be made for proposing that a collection of PAH cations is responsible for the DIB absorption. However, there is plainly more scope for future laboratory experiments, as well as theoretical work, before the hypothesis can be submitted to more rigorous tests.

## 10.7 SUMMARY

Many of the results reported in this chapter remain tentative and more experimental work is required to test some of the hypotheses. At the same time these experiments represent some of the most imaginative and far-reaching work which has been carried out on short-lived molecules. Thus modelling of atmospheric and interstellar chemical processes has been helped by the availability of spectroscopic data collected in the laboratory. This point is no better illustrated than by the matrix-isolation experiments which have help to interpret the processes occurring in interstellar grains. Indeed these matrix experiments have implications which extend as far as the possible mechanims of comet formation!

Other work has helped to explain the puzzling phenomena of unidentified infrared emission (UIR) in space, and the diffuse interstellar absorption bands (DIB). It can be seen that studies of short-lived molecules have applications which extend well beyond the confines of the research laboratory.

## REFERENCES

[1a] D. E. Milligan and M. E. Jacox, *J. Chem. Phys.* (1963), **39**, 712.

[1b] R. J. Saykally, P. G. Szanto, T. G. Anderson and R. C. Woods, *Astrophys. J.* (1976), **204**, L143.

[2] J. P. Burrows, D. W. T. Griffith, G. K. Moortgat and G. S. Tyndall, *J. Phys. Chem.* (1985), **89**, 266.

[3] D. W. T. Griffith and G. Schuster, *J. Atmos. Chem.* (1987), **5**, 59.

[4] J. M. Greenberg, *Sc. Am.* (1984), **250(6)**, 96.

[5] X. Tielens and L. J. Allamandola, *J. Phys. Chem.* (1983), **87**, 4220; J. M. Greenberg, C. E. P. M. van de Bult and L. J. Allamandola, *J. Phys. Chem.* (1983), **87**, 4243.

[6] J. H. Lacy, F. Baas, L. J. Allamandola, S. E. Persson, P. J. McGregor, C. J. Lonsdale, T. Geballe and C. E. P. M. van de Bult, *Astrophys. J.* (1984), **276**, 533.

[7] M. F. A'Hearn, P. D. Feldman and D. G. Schleicher, *Astrophys. J.* (1983), **274**, L99.

[8] M. Cohen, A. G. G. M. Tielens and L. J. Allamandola, *Astrophys. J.* (1985), **299**, L93; M. Cohen, L. J. Allamandola, A. G. G. M. Tielens, J. Bregman, J. P. Simpson, F. C. Wittebarn, D. Wooden and D. Rank, *Astrophys. J.* (1986), **302**, 737.

[9] W. W,. Duley and D. A. Williams, *Mon. Not. R. Astron. Soc.* (1981), **196**, 269.

[10] L. J. Allamondola, A. G. G. M. Tielens and J. R. Barker, *Astrophys. J.* (1985), **290**, L25.

[11] G. P. van der Zwet and L. J. Allamandola, *Astron. Astrophys.* (1985), **146**, 76.

[12] T. Shida and S. Iwata, *J. Am. Chem. Soc.* (1973), **95**, 3473.

[13] M. K. Crawford, A. G. G. M. Tielens and L. J. Allamandola, *Astrophys. J.* (1985), **293**, L45.

# Index

*ab initio* calculation, 137, 141, 161
acetaldehyde (Ch₃CHO), 184
acetic acid (CH₃COOH), 184
Acetone ((CH₃)₂CO), 184
acetylene (HC≡CH), 30–31, 74
alkane glass, 33, 35, 49, 102
alkane radical cation, 157–158
alkene complexes, 51, 72–76
alkene dimerization, 75
alkene epoxidation, 48
alkene hydrogenation, 75
alkene isomerization, 50
alkene metathesis, 79
alkene ozonolysis, 14, 124
alkene polymerization, 139
alkyne complexes, 72–76
aluminium, 175
aluminium–acetylene complex, 74
aluminium carbonyls, 67
aluminium–ethylene complex, 74
aluminium monobromide, AlBr, 136
aluminium monochloride, AlCl, 136–139
aluminium monofluoride, AlF, 137
aluminium monofluoride cation, AlF⁺, 158
aluminium monohydride AlOF, 137
aluminium thiochloride AlSCl, 138
aluminium trichloride, AlCl₃, 147
amino radical (NH₂), 161, 184
ammonia (NH₃), 171, 173
ammonia cation (NH₃⁺), 161
amorphous hydrogenated silicon, 93, 101, 153, 164
anion traps, 156, 164
ArF excimer laser, 100
argon ion laser, 103
argon matrix, 35, 40, 44, 52, 72, 102, 106, 118, 123, 136, 164
atmospheric pollutants, 182–184

benzyne, 126
bicyclo-[2.1.0]-pentane cation, 158
bismuth dimer, Bi₂, 83–85
Born–Haber cycle, 140
boron halide clusters, 153, 165–167

boron monochloride, BCl, 165–166
boron nitride, BN, 173
boron trichloride, BCl₃, 165

cadmium, 175
cadmium telluride, 172–173
caesium chlorate, CsClO₃, 143
caesium hexafluorouranate (V), CsUF₆, 147–148
caesium ozonide, CsO₃, 69
caesium peroxide, CsO₂, 69
calcium dimer, Ca₂, 82
carbenes, 121 *et seq.*
carbon monoxide cation, CO⁺, 156, 160
carbonyl oxides 14, 124–126
C–C bond activation, 76–78
C–H bond activation, 53, 76–78, 88, 123
chemical trapping, 25–26, 102–106, 110, 114
chemical vapour deposition (CVD), 93, 170–175
chemical vapour transport, 170
chlorine monoxide, ClO, 181–182
chlorine nitrate, ClONO₂, 181–182
chromium acetylacetonate, 174
chromium dimer, Cr₂, 82–83
chromium dioxide, CrO₂, 47, 71
chromium hexacarbonyl, Cr(CO)₆, 28 *et seq.*
chromium pentacarbonyl, Cr(CO)₅
    matrix-isolation studies, 32, 34–39, 66
    structure, 35
    reaction with CO, 37
    reaction with H₂, 54
    gas phase studies, 58
    anion, 59
    reaction of anion with O₂, 59
    formation from chromium atoms, 66
chromium tetracarbonyl, Cr(CO)₄, 58
chromyl carbonyl, CrO₂(CO)₂, 44–45
cobalt dichloride, CoCl₂, 146–147
cobalt porphyrins, 149
cobalt tetracarbonyl, Co(CO)₄, 66, 71
comets, 187
copper oxides, 71
copper trimer, Cu₃, 86
coronene, 188
cyclobutadiene, 127–129

cycloheptatriene cation, $C_7H_8^+$, 163–164
cyclohexane, 28
cyclopentadienylidene, 125–126

decarbonylation, 126
diacetyl peroxide, 153
1,4-dialumina-2,5-cyclohexadiene, 138
diazomethane, $CH_2N_2$, 79, 82, 121, 122
diboron tetrabromide, $B_2Br_4$, 166
diboron tetrachloride, $B_2Cl_4$, 165–167
dibromomethylene, $CBr_2$, 123
dichorobenzene cations, $C_6H_4Cl_2^+$, 162
dichloromethane, 164
dichromium decacarbonyl, $Cr_2(CO)_{10}$, 59
diethyl telluride, $(C_2H_5)_2Te$, 172–173
diffuse interstellar bands (DIB), 187–190
$m$-difluorobenzene cation, $m$-$C_6H_4F_2^+$, 160
dimanganese decacarbonyl, $Mn_2(CO)_{10}$, 48–50
dimanganese dicarbonyl, $Mn_2(CO)_2$, 66
dimanganese monocarbonyl, $Mn_2(CO)$, 66
dimethyl cadmium, $(CH_3)_2Cd$, 172–173, 175–176
dimethyl mercury, $(CH_3)_2Hg$, 82
dimethyl silylene $(CH_3)_2Si$, 93, 102–105, 108, 110–111, 118
dimethyl sulphoxide $(CH_3)_2SO$, 114
dimethyl zinc, $(CH_3)_2Zn$, 175–176
dinitrogen cation, $N_2^+$, 161
dinitrogen tetroxide, $N_2O_4$, 157
$trans$-dioxotetracarbonyl molybdenum, $trans$-$O_2Mo(CO)_4$, 45–46
$trans$-dioxotetracarbonyl tungsten, $trans$-$O_2W(CO)_4$, 45–46
diphenyl silylene, $(C_6H_5)_2Si$, 111
dirhenium decacarbonyl, $Re_2(CO)_{10}$, 48–50
disilacyclobutane, 116
disulphur, $S_2$, 187
disulphur dinitride, $S_2N_2$, 176–179

electric deflection, 18
electron diffraction, 20, 132, 141–142, 145–147
electron spin resonance (esr), 18, 68, 72, 85, 122, 148, 154, 156–158, 161–162, 177
ethylene, $C_2H_4$, 30–31, 51–53, 72–76
ethyl radical, $C_2H_5$, 172–173
ethynyl radical, $C_2H$, 181
extended X-ray absorption fine structure (EXAFS), 87

flames, 15, 16
flash photolysis, 14, 20–22, 26, 56, 103, 118, 126
nanosecond, 21
of metal carbonyls, 32–33
of silicon compounds, 103
flow techniques, 14, 18–20, 153
fluorine anion, $F_2^-$, 156
fluorine atom, F, 17, 155
fluorine dioxide, $FO_2$, 18
formaldehyde, HCHO, 48, 184
formic acid, HCOOH, 184
formyl radical, HCO, 184
free radicals, 81–82, 153–167
freon, 181

glass, 157–158, 188

gallium arsenide, GaAs, 171, 175
gallium nitride, GaN, 173–174
gamma rays, 18
gas chromatography, 102

halobenzene cations, 158–163
heptaboron heptabromide, $B_7Br_7$, 166–167
hexafluorobenzene cation, $C_6F_6^+$, 158–159
high-temperature molecules, 15, 132 $et\ seq.$
hydrogen isocyanide, HNC, 181
hydrogen sulphide, $H_2S$, 187
hydroxyl radical, OH, 16
hydroxymethyl radical, $CH_2OH$, 155–156

indium phosphide, InP, 171–172
interstellar grains, 184–187
intracavity laser absorption spectroscopy, 100
IRAS–Araki–Alcock 1983 VII (comet), 187
iridium tricarbonyl, $Ir(CO)_3$, 127
iron clusters, 87–88
iron dichloride, $FeCl_2$, 146–147
iron-ethylene complexes, 75
iron-methanol adduct, $Fe.CH_3OH$, 79
iron methylene, $Fe=CH_2$, 79
iron monohydride, FeH, 181
iron nitrosyls, 51
iron pentacarbonyl, $Fe(CO)_5$, 28, 33, 39–41
iron porphyrins, 149
iron tetracarbonyl, $Fe(CO)_4$
    structure, 39
    electronic configuration, 39–40
    laser induced isomerization, 40
        reaction with $H_2$, 54
isolabal relationship, 124, 127
isomerization
    of $Fe(CO)_4$, 40
    of silylenes, 93, 117–118
    of tetra-$tert$-butyl tetrahedrane, 130
    of dichlorobenzene cations, 162
    of $C_7H_8^+$ cations, 163

Jahn–Teller distortion, 157–159

ketene, LH2C=CO, 126
kinetic studies, 26, 54, 58, 68, 93, 95–96, 107–108, 126, 175–176
Knudsen cell, 64
krypton (liquid), 54
krypton matrix, 133

laser induced fluorescence (LIF), 82, 100
laser magnetic resonance (LMR), 18, 156
laser writing, 175–176
lead dimer, $Pb_2$, 83–85
lead mirror, 154
liquid phase epitaxy, 170
lithium heptamer, $Li_7$, 85
lithium nitrate, $LiNO_3$, 143–144
lithium peroxide, $LiO_2$, 68
lithium trimer, $Li_3$, 85

magnesium dimer, $Mg_2$, 82
manganese carbonyl, $Mn_2(CO)_{10}$, 48–50
manganese heptoxide, $Mn_2O_7$, 50
manganese nitrosyls, 51
manganese oxides, 71
manganese pentacarbonyl $Mn(CO)_5$, 66
manganese porphyrins, 149
mass spectrometry, 18, 59, 132, 150, 164–167, 172, 177
matrix isolation
  general, 14, 22–26, 76–81
  of metal carbonyls, 34–50, 54–59, 64–68, 71–72, 89
  of metal nitrosyls, 51
  of metal oxides, 47–48, 50, 68–72, 145–147
  of metal–alkene complexes, 52–53, 72–75
  of metal–alkyne complexes, 72
  of free radicals, 81–82
  of metal dimers and clusters, 83–88
  preparative, 89
  of silicon compounds, 93, 97–99, 102–106, 118
  of carbenes, 122–123
  of cyclobutadiene, 128–129
  of high-temperature molecules, 132–150
  of metal oxysalts, 142–145
  of radicals and ions, 153–167
  of sulphur nitrides, 176–179
  of atomspheric trace gases, 182–184
  modelling interstellar proceses, 184–190
McLafferty rearrangement, 164
mercury, 172
mercury discharge, 165
mercury telluride, HgTe, 172–173
metal–alkene complexes, 51, 72–76
metal alkoxides, 174
metal–alkyne complexes, 72–76
metal atoms, 64 *et seq.*
metal carbonyls, 17, 28 *et seq.*, 64–68, 71, 89, 124
  gas phase, 58–60
  formation from metal atoms, 64–68, 89
  isolobal relationship with carbenes, 124
metal chlorates, 142–143
metal chlorides, 132
metal clusters, 85–88
metal-dihydrogen complexes, 54–55
metal dimers, 82–85
metal hydrides, 54–55
metal nitrides, 173–174
metal nitrosyls, 51–53
metal organic chemical vapour deposition (MOCVD), 171–175
metal oxides, 47, 68–72, 132, 173–174
metal perrhenates, 142–143
metal phosphates, 143
metal phosphites, 143
metal silicides, 174–175
metal vapour synthesis (MVS), 88
metal–water reactions, 78–79
methane matrix, 35, 36, 41, 51, 155
methanol, $CH_3OH$, 48, 79, 155,–156
methine, CH, 127, 155
methoxyl radical, $CH_3O$, 18, 155–156

methyl bromide, $CH_3Br$, 176
methyl cyclohexadiene cation, $C_7H_8^{'}$, 164
methylene, $CH_2$, 109, 121–123, 126
methyl iodide, 81
methyl metal hydrides, 76–78
methyl radical, $CH_3$, 18, 81, 154–155
methyl silene, $CH_3Si(H)=CH_2$, 103
microwave discharge, 101, 174
molecular beam epitaxy (MBE), 170
moelcular ions, 153 *et seq.*
molybdenum dimer, $Mo_2$, 82
molybdenum dioxide, $MoO_2$, 47, 145
molybdenum hexacarbonyl, $Mo(CO)_6$
  photochemistry, 33–34
  photooxidation, 44–48
molybdenum pentacarbonyl, $Mo(CO)_5$
  matrix isolation studies, 34–39
  reaction with CO, 37
  reaction with $N_2$, 42–43, 54
molybdenum–silica heterogeneous catalyst, 48
molybdenum tetracarbonyl, $Mo(CO)_4$, 41–42
molybdenum tricarbonyl, $Mo(CO)_3$, 41–42
molybdenum trioxide, $MoO_3$, 47, 145
monocarbonyl gold peroxyformate $Au(CO)(CO_3)$, 71–72

neon matrix, 36, 156
Nernst glower, 40
neutron irradiation, 97
nickel carbonyls, 64–65, 136
nickel dichloride, $NiCl_2$, 146–147
nickel–ethylene complex, 72–73
nickel oxides, 71
nickel tetracarbonyl, $Ni(CO)_4$, 54, 64–65
nitrogen dioxide, $NO_2$, 181–182
nitrogen matrix, 132–133
nitrous acid, $HNO_2$, 184
nobel gases (liquid), 53–56
nonaboron nonachloride, $B_9Cl_9$, 166
non-Berry pseudorotation, 41

Orion nebula, 181
oxetane, 108
oxygen atoms, 48
oxysilirane, 114
ozone, 69, 181–182
ozonolysis, 14, 124

perchlorate salts, 15
phosphorus dioxychloride, $PO_2Cl$, 132, 141
phosphorus mononitride, PN, 133–134
phosphorus oxychloride, $POCl$, 15, 20, 132, 139–141
phosphorus oxyfluoride, POF, 140
phosphorus oxytrichloride, $POCl_3$, 15, 20, 139
phosphorus thiochloride, PSCl, 140
photochemical etching, 176
photolysis, 15, 17, 28 *et seq.*, 76–81, 94–95, 98–105, 114, 121–131, 153–164, 172–176, 182, 184–190
plasma, 17, 164–167, 174
plasma etching process, 103, 107

polarization studies
  on $Cr(CO)_5$, 37
  on $Mo(CO)_5N_2$, 43
  on $Re_2(CO)_9$, 49
  on $(CH_3)_2Si$, 103
polyacetylene, 181
polycyclic aromatic hydrocarbon (PAH), 186–190
polythiazyl, $(SN)_x$, 176–179
potassium chloride, KCl, 145
potassium trimer, $K_3$, 85–86
pyrolysis, 82, 94–95

Raman spectroscopy, 42, 68
rhenium atoms, 88
rhenium carbonyl, $Re_2(CO)_{10}$, 48–50
rhodium monocarbonyl, $Rh(CO)$, 136
rhodium nitrosyls, 51
ruthenium pentacarbonyl, $Ru(CO)_5$, 89

semiconductor, 170–175
shock tube, 15, 16, 95
silacyclopropene, 112–113
silanes, 94, 171
  pyrolysis of, 94–95
  photolysis of, 94–95, 98, 100, 102, 105, 114
silanones, 114
silicon, 93, 101, 136, 175
silicon atoms, 94, 97, 99
silicon dicarbonyl, $Si(CO)_2$, 97
silicon dichloride, $SiCl_2$, 106
silicon difluoride, $SiF_2$, 97, 106–107
silicon dihalides, 105–107
silicon–dinitrogen complex, $Si(N_2)$, 97
silicon dioxide, $SiO_2$, 134, 175
silicon monocarbonyl, $Si(CO)$, 97
silicon monoxide, SiO, 15, 132–136
silicon monoxide cation, $SiO^+$, 158
silicon monosulphide, SiS, 133
silicon oxydichloride, $SiOCl_2$, 135
silver–silicon monoxide complex, $Ag(SiO)$,
  135–136
silver tricarbonyl, $Ag(CO)_3$, 68
silver trimer, $Ag_3$, 86
silylenes, 93 *et seq.*, 123, 171
  production of 94–99, 102, 106
  insertion reactions of, 107–110
  addition reactions of, 110–114
  abstraction reactions of, 114
  dimerization reactions of, 114–116
  isomerization reactions of, 116–118
  formation of adducts with O-donors, 108,
    113–114
  in plasmas, 164–1654
silyl radical, $SiH_3$, 96, 164–165
sodium chloride, NaCl, 145
sodium ozonide, $NaO_3$, 69
sodium peroxide, $NaO_2$, 68–69
sodium phosphate, $NaPO_3$, 15
sodium phosphite, $NaPO_2$, 15
sodium trimer, $Na_3$, 85
sputtering, 15
stannane, $SnH_4$, 157

stratosphere, 181–182
supersonic molecular beam, 20, 87–88

tetrahedrane, 127–129
tetrahydrofuran, 109
tetrairidium dodecacarbonyl, $Ir_4(CO)_{12}$, 127
tetramethyl lead, $(CH_3)_4Pb$, 154
tetranitrogen cation, $N_4^+$, 161
tetrasulphur tetranitride, $S_4N_4$, 176–178
tetra-*tert*-butyl tetrahedrane, 130
thallium nitrate, $TlNO_3$, 142
thallium oxide, $Tl_2O$, 145
thallium perrhenate, $TlReO_4$, 142
thiazyl radical, SN, 177–178
thionylimide, HNSO, 178
time resolved infrared spectroscopy, 56–58
tin carbonyls, 67
tin oxides, 70
titanium monoxide, TiO, 181
titanium nitride, TiN, 174
titanium tetrachloride, $TiCl_4$, 174
toluene cation, $C_7H_8^+$, 163–164
transistor, 170
1,3,5-trifluorobenzene cation, $C_6H_3F_3^+$, 158–159
triiron dodecacarbonyl, $Fe_3(CO)_{12}$, 50
trimethyl aluminium $(CH_3)_3Al$, 175–176
trimethyl gallium, $(CH_3)_3Ga$, 173
triruthenium dodecacarbonyl, $Ru_3(CO)_{12}$, 50, 89
tungsten, 175
tungsten dioxide, $WO_2$, 16
tungsten hexacarbonyl, $W(CO)_6$, 28, 44–48
tungsten hexafluoride, $WF_6$, 174
tungstenocene, $Cp_2W$, 77
tungsten pentacarbonyl, $W(CO)_5$, 34–39
  reaction with $C_2H_2$ or $C_2H_4$, 30–31
  matrix isolation studies, 34–39
  reaction with CO, 37
tungsten silicide, 174
tungsten trioxide, $WO_3$, 47

unidentified infrared emission (UIR), 187–190
uranium carbonyls, 67
uranium dioxide, $UO_2$, 71
uranium hexacarbonyl, $U(CO)_6$, 66–67
uranium trioxide, $UO_3$, 71

vacuum ultraviolet radiation, 17, 82, 98, 116, 155,
  158, 186
vanadium nitrosyls, 51

W33A (astronomical object), 186–187
water, $H_2O$, 78–79
water cation, $H_2O^+$, 156

zenon (liquid), 15, 25, 54
zenon matrix, 36, 122
X-rays, 17, 155

Zeise's salt, 139
Ziegler–Natta catalyst, 139
zinc, 175